# Mobile Technology and Academic Libraries:

## *Innovative Services for Research and Learning*

edited by Robin Canuel and Chad Crichton

Association of College and Research Libraries
A division of the American Library Association
Chicago, Illinois 2017

The paper used in this publication meets the minimum requirements of American National Standard for Information Sciences–Permanence of Paper for Printed Library Materials, ANSI Z39.48-1992. ∞

Cataloging-in-Publication data is on file with the Library of Congress.

Copyright ©2017 by the Association of College and Research Libraries.

All rights reserved except those which may be granted by Sections 107 and 108 of the Copyright Revision Act of 1976.

Printed in the United States of America.

21  20  19  18  17    5  4  3  2  1

# Table of Contents

ix .......... Acknowledgments

xi .......... Introduction
   *Robin Canuel and Chad Crichton*

1 .......... **Chapter 1. The Mobile Context: A User-Centered Approach to Mobile Strategy for Libraries**
   *Edward Bilodeau*

   - 1 ....... Introduction
   - 2 ....... Mobile Technologies and the Decline of the Desktop Computing Paradigm
   - 4 ....... The Mobile Context
   - 9 ....... Exploring the Mobile Context of Your Users
   - 11 ..... Informing Your Mobile Technology Strategy
   - 13 ..... Conclusion
   - 13 ..... References

15 ......... **Chapter 2. The Development of an Academic Library's Mobile Website**
   *Junior Tidal*

   - 15 ..... Introduction
   - 15 ..... Literature Review
   - 16 ..... Background
   - 17 ..... History of the City Tech Library Mobile Site
   - 30 ..... Future Improvements
   - 32 ..... References

35 ........ **Chapter 3. A Mobile-First Library Site Redesign: How Designing for Mobile Provides a Better User Experience for All**
   *Nathan E. Carlson, Alec Sonsteby, and Jennifer DeJonghe*

   - 35 ..... Introduction
   - 36 ..... The Mobile Landscape
   - 38 ..... Understanding User Needs and Behaviors
   - 41 ..... Pre-design Phase
   - 42 ..... Site Design and Testing
   - 47 ..... Challenges and Future Directions
   - 48 ..... Acknowledgments
   - 49 ..... Appendix 3A. Metropolitan State Library Homepage Redesign, User Survey
   - 51 ..... Appendix 3B. Metropolitan State Library Homepage Redesign, Test Script: Wireframe 2
   - 52 ..... References

**55** ....... **Chapter 4. Selfie as Guide: Using Mobile Devices to Promote Active Learning and Student Engagement**
*Sarah LeMire, Stacy Gilbert, Stephanie Graves, and Tiana Faultry-Okonkwo*

    55 ..... Introduction
    56 ..... Library Tours as Academic Library Outreach
    56 ..... Active Learning and Mobile Technology
    58 ..... Mobile Technology and Library Tours
    60 ..... Developing the Selfie-Guided Tour
    65 ..... Lessons Learned and Future Directions
    67 ..... Conclusion
    68 ..... Appendix 4A. Gateway Alignment Grid
    69 ..... References

**73** ....... **Chapter 5. Beyond Passive Learning: Utilizing Active Learning Tools for Engagement, Reflection, and Creation**
*Teresa E. Maceira and Danitta A. Wong*

    73 ..... Introduction
    74 ..... Literature Review
    78 ..... Tools and Learning Activities
    85 ..... Apps Workshops
    86 ..... Technical Issues
    87 ..... Conclusion
    88 ..... References

**91** ....... **Chapter 6. Getting Meta with Marlon: Integrating Mobile Technology into Information Literacy Instruction**
*Regina Lee Roberts and Mattie Taormina*

    91 ..... Introduction
    92 ..... Background
    94 ..... Case Study: The Marlon Riggs Collection
    96 ..... Mobile Device Selection
    98 ..... Reflections
    100 ... Conclusion
    101 ... Appendix 6A. Workshop Stations
    103 ... References

**105** ..... **Chapter 7. Clinical Resources for the Digital Physician: Case Study and Discussion of Teaching Mobile Technology to Undergraduate Medical Students**
*Maureen (Molly) Knapp*

    105 ... Introduction
    107 ... Course Design
    107 ... Audience Response and Feedback

108 ... Readings and Discussion
118 ... How Does This Tie into the Information Literacy Framework?
119 ... Results and Conclusion
120 ... Appendix 7A. List of Subscription and Free Apps Reviewed in Clinical Resources for the Digital Physician App Demonstration
121 ... References

**123 ...... Chapter 8. Mobile Technology Support for Field Research**
*Wayne Johnston*

123 ... Preamble
124 ... Introduction
125 ... Literature Review and Methodology
127 ... Open Data Kit (ODK)
129 ... Deployments
130 ... Knowledge Base
132 ... Conclusions
133 ... References

**135 ...... Chapter 9. From Start to Finish: Mobile Tools to Assist Librarian Researchers**
*Mê-Linh Lê*

135 ... Introduction
136 ... A History of Apps
136 ... The Librarian as Researcher
140 ... Mobile Apps for Research
149 ... Apps in Action
150 ... Conclusion and Future Directions
150 ... References

**153 ...... Chapter 10. A Novel Application: Using Mobile Technology to Connect Physical and Virtual Reference Collections**
*Hailie D. Posey*

153 ... Introduction
156 ... Research Guides and the User Experience
159 ... Building the Theology Collections Portal in Scalar
166 ... Strengths, Shortcomings, and Future Directions for the Theology Collections Portal
168 ... References

**171 ....... Chapter 11. Adding Apps to Our Collections: A Pilot Project**
*Willie Miller, Yoo Young Lee, and Caitlin Pike*

171 ... Introduction
172 ... Literature Review
174.... Project Background
174.... App Selection

175 ... Instruction with iPads
179 ... Barriers or Issues
180 ... Conclusion
181 ... References

**183 ...... Chapter 12. Tablets on the Floor: A Peer-to-Peer Roaming Service at Atkins Library**
*Barry Falls, Beth Martin, and Abby Moore*

183 ... Introduction
184 ... Logistics: Making It Happen
185 ... Pilot Project
190 ... Implementation
191 ... The Student Library Advisory Board's Opinion on Roamers
191 ... Focus Groups
192 ... Marketing
192 ... Hours of Operation
193 ... Unexpected Benefits
194 ... Reflection and Next Steps
195 ... References

**197 ....... Chapter 13. Using Proximity Beacons and the Physical Web to Promote Library Research and Instructional Services**
*Jordan M. Nielsen and Keven M. Jeffery*

197 ... Introduction
199 ... Background
200 ... Implementation
204 ... Promotion
204 ... Lessons Learned
205 ... Technical Hurdles
205 ... Next Steps
207 ... References

**209 ...... Chapter 14. Gamification Using Mobile Technology in the Classroom: A Positive Benchmark for the Future of Higher Education**
*Avery Le*

209 ... Introduction
211 ... Gamification in the Classroom
215 ... The Benefits of Gamification
216 ... Possible Drawbacks
219 ... Solutions for Drawbacks
219 ... Conclusion
220 ... References

**223 ...... Chapter 15: Bringing Texts to Life: An Augmented Reality Application for Supporting the Development of Information Literacy Skills**
*Yusuke Ishimura and Martin Masek*

- 223 ... The Undergraduate Student's Experience of Writing Research Papers
- 224 ... What Is Augmented Reality?
- 225 ... Development of the AR Application: The Trailblazer Project and Text Recognition
- 227 ... Content Development
- 230 ... Results
- 234 ... Conclusion and Future Direction
- 235 ... Acknowledgment
- 235 ... References

**237 ...... Chapter 16. Virtual Reality Library Environments**
*Jim Hahn*

- 237 ... Introduction
- 238 ... General Virtual Reality Hardware
- 238 ... Contemporary Virtual Reality Hardware
- 239 ... Google Cardboard Virtual Reality Experience
- 240 ... Review of Virtual Reality Applications and Current Academic Uses
- 241 ... Library Virtual Reality Use Cases for Research and Teaching
- 245 ... Developer Resources
- 246 ... Future Directions
- 247 ... References
- 248 ... Works Consulted

**249 ...... Chapter 17. Wearable Technologies in Academic Libraries: Fact, Fiction and the Future**
*Ayyoub Ajmi and Michael J. Robak*

- 249 ... Introduction
- 250 ... Market Analysis
- 251 ... Google Glass Explorer
- 256 ... Other Smart Glasses
- 256 ... Activity Trackers
- 257 ... Smartwatches
- 257 ... Other Wearable Devices
- 258 ... Wearables in Academic Libraries
- 260 ... Challenges
- 261 ... Conclusion
- 262 ... References

**About the Authors**

# Acknowledgments

The editors would like to acknowledge the assistance of the following individuals and institutions that provided us with support and encouragement throughout the course of this project: the McGill University and University of Toronto Scarborough Libraries, Kathryn Deiss, Erin Nevius, and the Association of College and Research Libraries. We would like to thank the following individuals for reading various drafts of chapters and providing their thoughtful comments and feedback: Edward Bilodeau, Eamon Duffy, Dawn McKinnon, Lonnie Weatherby, and Deena Yanofsky. We would also like to thank the chapter authors for sharing their innovative and interesting ideas and services and for writing such high-quality chapters, and a special thank you to our families for their steadfast support and understanding—Jennifer, Matthew and Nicholas, and Julie.

# Introduction

## Robin Canuel and Chad Crichton

As we approach the tenth anniversary of the introduction of the Apple iPhone, mobile technology has become a ubiquitous presence in the lives of today's students and faculty. The maturing of this technology has led to our becoming more and more comfortable in a world where digital information flows seamlessly from screen to screen as we move about our daily lives, freeing us somewhat from the constraints of wired technology. This evolution presents both risks and opportunities for academic librarians, operating as we do in a field that is both uniquely tied to a static sense of "place" in the public imagination and at the same time passionately devoted to the freedom, spread, and accessibility of information for the public at large. The following chapters explore the responses of academic libraries to this maturing of mobile technology, as librarians around the world work to adapt their spaces, collections, teaching, and services to the new possibilities presented by mobile technology. In libraries today, no longer do we expect, nor require, our users to have a physical presence in the library. Leveraging the potential of smartphones, tablets, and even wearable technologies allows academic librarians to further expand their reach to students and faculty beyond the library's walls. Furthermore, by understanding how mobile technology changes the behavior of our users, we can gain new insights into their needs and make improvements to our traditional services and spaces in order to better contribute to faculty research and student learning.

The first step in embracing the challenges and opportunities of mobile technology is understanding the context in which library users employ this technology and adapting our online presence to respond not only to the limitations of mobile devices, but also to their ability to open up new possibilities for users. In chapter 1, Edward Bilodeau explains this "mobile context" and how it has significantly altered the ways in which people interact with technology. Bilodeau highlights the importance of this changed context for web designers and also for academic librarians thinking about integrating mobile resources and devices into their practice. Junior Tidal follows with a practical case study of the design of a mobile website through a number of iterations, using a wide variety of approaches and tools. Tidal presents a historical review of the work done at his institution in response to the evolving needs of mobile library users over the years as an overview of how mobile websites have evolved to respond to changes in both technology and user behavior. Chapter 3, from Carlson, Sonsteby, and DeJonghe, highlights the importance of collecting insights from users in order to ensure that a library's mobile-inspired website redesign meets the needs of today's faculty and students. This user-centered design approach incorporates feedback comprised of real-world user experiences with mobile websites in order to improve its organization and functionality in line with user expectations.

Subsequent to the current response of librarians to the new mobile context and the online design consequences that stem from this new reality, our focus shifts to the implications of this technology for library orientation and instruction work. In the time-honored realm of library orientation, LeMire, Gilbert, Graves, and Faultry-Okonkwo describe their development of new, self-guided library tours that leverage mobile devices and that can be integrated into information literacy instruction sessions. In our fifth chapter, Maceira and Wong discuss the integration of iPads into in-class information literacy sessions and detail the advantages that mobile technology can bring to the classroom. Roberts and Taormina continue our coverage of mobile technology for teaching in chapter 6 with their case study of its integration into their workshop, introducing the Special Collections and Archives of Stanford University to anthropology students. Their use of iPads as recording devices in the classroom introduces their students to the use of mobile technology as a tool for research in the field, a topic covered in more detail in the following chapters. Maureen Knapp at Tulane University, meanwhile, focuses on the importance of mobile technology to the health care professions in a potentially less remote real-world context. Knapp's instruction work is focused on medical students, with whom she discusses the advantages of leveraging mobile technology to provide improved bedside patient care utilizing specialized mobile apps and custom hardware. Her workshops introduce these students to some of the ways in which mobile technology can facilitate their work in the field, and its use for fieldwork continues as a theme in our succeeding chapters.

Wayne Johnston begins our discussion of mobile technology as a research tool with his chapter covering some of the unique ways in which mobile technology can assist university researchers engaged in fieldwork around the world. Johnston's personal experiences and numerous discussions with field researchers in a number of disciplines result in several colorful anecdotes that drive home mobile technology's potential to have an immensely positive impact on the work of researchers working to gather data out in the "real world," away from the confines, and infrastructure, of their home institutions. In chapter 9, Mê-Linh Lê brings the focus back to the library explicitly, and she tackles the research implications of mobile technology from the perspective of librarians engaging in their own research. Lê covers the use of mobile technology for all aspects of the research process, from data collection through publication and dissemination, highlighting and suggesting some of the best apps for librarians when conducting their research.

Earlier chapters having covered mobile technology from the perspectives of web design and teaching, Hailie Posey's chapter then introduces us to the use of mobile technology from the perspective of collections work. With her case study on the deployment of iPad kiosks into the physical stacks, she demonstrates how patrons who are browsing the stacks can have rapid and convenient access to contextual digital information that can be used to supplement these print collections. Posey's chapter illustrates the ways in which mobile technology can be combined with traditional collections to enhance the user's browsing experience and his or her ability to engage more deeply with print collections. Miller, Lee, and Pike also discuss mobile apps from a collections perspective, with the apps themselves being selected for addition to the collection in a manner similar to traditional print and electronic materials. Miller and his colleagues discuss their experiences and conclusions with regard to how best to collect, curate, deploy, and promote apps as part of our campus collections.

Outside of the classroom and the stacks, mobile technology is also impacting that most central service in academic libraries—reference and research assistance. Chapter 12, by Falls, Martin, and Moore, discusses the use of mobile technology to provide a peer-to-peer roaming reference service for students and researchers at University of North Carolina at Charlotte. Mobile devices can be deployed in order to free reference staff from the confines of the traditional reference desk and enable them to meet with students and faculty at their point of need. Moore and her coauthors detail the promise and pitfalls of implementing a roving reference service, including both unexpected barriers and unanticipated benefits. In the following chapter, Nielsen and Jeffery describe the next step in mobile reference support, whereby students and faculty can have contextual information about the collections with which they are interacting (or any space with which they are interacting, for that matter) beamed directly to their mobile devices without the direct intervention of library staff in the moment. Nielsen and Jeffery's library implemented proximity beacons in various locations in and around their library to work in conjunction with mobile devices to connect the physical world with the virtual and to assist patrons in navigating collections and services. With the proper app installed, library users within physical proximity of one of these stationary beacons can have online resources "pushed" to their devices—resources that provide contextual information to the users tailored to their physical location. This leveraging of mobile technology to provide contextual, unmediated, location-based information and services to users when and where they are is a very new development, and it demonstrates the potential impact that new uses of mobile technology could have in the future.

In "Gamification Using Mobile Technology in the Classroom," Avery Le discusses how mobile technology can be used to leverage the appeal of online gaming in order to foster better learning experiences. This "gamification" of the classroom promises many potential benefits to the instructor looking to reach out to today's students and maintain their attention. In chapter 15, Yusuke Ishimura and Martin Masek explore the emerging world of augmented reality and the ways in which mobile devices can be used to superimpose digital information on the real world in order to enhance the understanding of researchers and students. The app described by Ishimura and Masek allows for information to be digitally superimposed on a physical text in order to enhance a print document in real time with added detail and context. Next, Jim Hahn at the University of Illinois at Urbana–Champaign takes us from augmented reality to virtual reality as he investigates how new advances in mobile hardware allow researchers and students to harness virtual environments for their research and learning in ways that were never before possible. Finally, Ayyoub Ajmi and Michael Robak take us beyond "traditional" mobile devices to introduce us to the world of "wearable" technology. Ajmi and Robak discuss their early adoption of Google Glass and explore the immense potential and still evolving challenges of the ultimate expression of "mobility" as they attempt to foresee what the future may hold in an emerging world in which technology is worn as much as carried.

As mobile technology continues to evolve, academic librarians will continue to look for new innovations that will bring not only new efficiencies, but also entirely new methodologies to researchers and students. The future promises continuous change as information surges beyond the boundaries of wired infrastructure and our ability to gather, manipulate, and interpret data in new ways (and in new places) expands. Our work as professional academic librarians will be to strive to anticipate the ways in which our notions of research, teaching, and learning—and the importance of location and context for those activities—will be challenged and changed by mobile technologies. Mobile devices are becoming consistently and continuously more powerful, while they simultaneously become less intrusive and more intuitive. In the future it will become necessary for academic librarians to be increasingly conscious of the importance of the mobile context when developing their collections, spaces, services and teaching. The continuous evolution of mobile technology provides new ways to connect faculty and students to the world of information available to them through their university's libraries, and it is incumbent upon academic librarians to embrace these changes and to encourage the development of collections, spaces, and services that fully harness the potential of our ever-present mobile devices, untethered from the wired world.

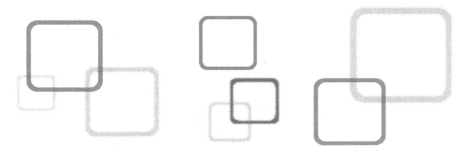

CHAPTER 1

# The Mobile Context
## A User-Centered Approach to Mobile Strategy for Libraries

*Edward Bilodeau*

## Introduction

Libraries have always looked for ways to make use of new technologies to enhance the resources and services that they provide to their user communities. As the use of mobile devices became more widespread, many libraries invested in developing or purchasing mobile applications and websites to make their resources and services more easily available on this new platform. The primary goal of these initiatives was to adapt existing features and content to mobile devices, adjusting to make use of the form factors and interaction models that were different from those offered by traditional workstation computers. While these projects typically succeed in providing functional mobile applications, in many cases patron engagement failed to meet the expectations set by the library at the beginning of the project.

Most libraries have adopted practices such as usability testing to support their design and development process, but this alone is not enough to realize the full potential offered by mobile devices. The approaches and assumptions underlying how we conceive, design, and implement technology are, for the most part, grounded in the traditions of desktop computing. However, we can no longer rely on the computing paradigm presented by desktop computing as an accurate or representative description of how people use and interact with their personal computing devices. The mobile context provides libraries with a perspective for understanding how their users make use of mobile devices and other forms of computing technology, enabling libraries to design and deploy mobile applications and services that meet the needs of their user communities.

## Mobile Technologies and the Decline of the Desktop Computing Paradigm

Today's mobile technologies have ushered in a new era in personal computing. People now have a wider choice of devices available to them to fulfill any task. They can also choose the device that best suits their needs and preferences. Although desktop computers are still used by many on a daily basis, they are no longer the only, or even the primary device of choice (Kim, 2013). This is an important shift, as our assumptions of how technology can be used to carry out or support tasks have largely been defined by our experience of computing "tethered to a desktop computer or laptop" (Bentley & Barrett, 2012, p. 17). Libraries need to move away from a technology-centered approach to application development and to recognize the decline of the desktop paradigm as a model for their users' computing experience.

Most software and web applications were (and still are) designed with the interface, input devices, and configuration of the typical desktop computer in mind. The assumption was that the user would conform to the needs of the technology in order to be able to make use of the application. In this model, the user sits in front of the computer screen with the input devices (typically they keyboard and mouse) in reach (see figure 1.1).

**FIGURE 1.1**
The desktop context.

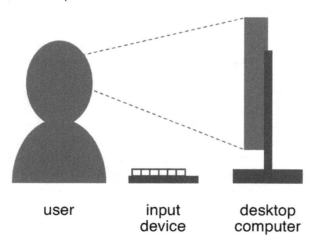

user     input device     desktop computer

The user was expected to adapt to the availability and realities of the technology. In order to use the application, the individual had no choice but to travel to wherever the desktop computer was located, sit before the computer, and focus on the task at hand. The user was also expected to undergo training to learn how to use the application in accordance with the developer's design.

In the desktop model of computing, the interaction between the user and the computer is considered to be independent of context, with the attention of the user entering "into the 'nonspace' of [the computer's] interface" (Greenfield, 2006, p. 71). Where this interaction is taking place, the events occurring around it, the exact positioning of the user in relation to the computer interface—none of this is considered relevant to the conceptualization or design of the application. It is worth noting that even in a recent review of the literature covering usability methods and models, it was found that fewer than 10 percent of the papers considered context as a relevant component of any model of usability (Harrison, Flood, & Duce, 2013).

Our approach to designing and developing computer applications has, for the most part, been grounded in the early days of mainframe computing. At that time, people, and entire organizations, had very little choice but to conform to technology, often using the computer in laboratory-like settings. With the birth of personal computing in the late 1970s, and its rapid acceleration in the 1980s, people began to have a choice of which computer to use, where to use it, and what applications to use. The physical reality of desktop computers did place some constraints on where and how they were used, with most people choosing to sit before them at a desk, much as they used the typewriters the computers were modeled after.

Personal computers were being used in offices, homes, schools, and other contexts that were already far removed from the controlled environments where mainframe workstations had been used. The 1980s also saw the introduction of portable computers, the first mobile computing technology. The first models were little more than desktop computers that were designed to be easier to move than a typical computer and monitor. Advances in technology and miniaturization led to the laptop form factor, giving users a personal computer that was truly portable. All of these devices, however, were for the most part considered to be variations or instances of the desktop computer. People used the same applications that they used on their desktop computers. People adjusted to the smaller keyboards and alternative pointing devices (trackballs, trackpads, etc.) that were created to support the traditional desktop interaction model.

The introduction and rapid adoption of smartphones and tablet computers over the past fifteen years has ushered in a new era in mobile computing, one that is forcing us to rethink the basic paradigm of personal computing. Modern mobile technologies have provided people with new ways of using computing technology without being tethered to a desktop computer. People will continue to use desktop computers and laptops, but will supplement these with smartphones, tablets, and new forms of personal mobile computing technologies such as wearable devices. In light of this new reality, the model provided by the traditional desktop context, that of a detached user placed before a screen, is not likely to provide us with the guidance and understanding that we need in order to provide our users with the technology choices that they want and need. We can no longer assume that the user will be at a desktop, nor can we say with any certainty which technology the user will employ to carry out a given task. We have to account for the ways in which the various elements of the users' context will influence their decisions of which technologies to use or how that context will impact their ability to make effective use of the technology at their disposal. If we want to provide the communities our libraries serve with useful and compelling online services and applications, we need to focus our efforts on understanding the mobile context of our users.

## The Mobile Context

Mobile context provides a rich perspective that includes the user, the technology, and everything that makes up their environment. In seeking to understand what constituted the mobile context, Hinman (2012) and her team asked ten people to take pictures whenever they used their phones. Upon analyzing the photos, they realized that the mobile context was anywhere and everywhere (Hinman, 2012).

There are two related but separate perspectives on the mobile context. The first considers the mobile context from the perspective of the technology, or device, while the second looks at it from the perspective of the person using the device. It is this second, user-centered perspective on the mobile context that is most relevant to the conception and design of mobile applications.

### *The Mobile Context of the Technology*

Designers and developers working in the field of context-aware computing have a conception of mobile context that is by necessity focused on the technology. Their goal is to define the context that needs to be sensed and "understood" by the application in order to enable functionality that leverages the mobile nature of the device. Unlike traditional computer and software development, the context with mobile devices is different because the device can potentially sense, react to, and interact with the world around it in ways that a desktop computer or laptop cannot (Bentley & Barrett, 2012). Writing about context-aware applications, Dey, Abowd, and Salber (2001) define context as the following: "Context: any information that can be used to characterize the situation of entities (i.e., whether a person, place, or object) that are considered relevant to the interaction between a user and an application, including the user and the application themselves. Context is typically the location, identity, and state of people, groups, and computational and physical objects" (p. 106).

This technology-driven perspective on the mobile context is important to libraries, as the capacity of mobile devices informs our thinking about what kinds of solutions are feasible. However, the capacities and limitations of the technology cannot be the focus of our thinking. Our goal is not to implement technology but to provide solutions that meet the needs of our users. In order to do that, our thinking needs to be grounded in an understanding of the user.

## *The Mobile Context of the User*

The purpose of the concept of the mobile context is to provide us with a framework to understand the possible elements that can have an impact on the users' choice, and use, of mobile technology to carry out tasks that move them towards the successful realization of their goals. The mobile context describes the constraints that need to be considered when designing, developing, and implementing mobile technologies. These constraints can be categorized as device constraints, environmental constraints, and human constraints (Hinman, 2012, p. 46). Device constraints refer to the limitations of the mobile devices' size, display, and input mechanisms. Environmental constraints consist of the physical, visual, and audible environment that the person is surrounded by and moving through. The human constraints include the person's cognitive ability, cultural influences, as well as his or her personal ergonomic limitations for working with mobile devices.

The diagram in figure 1.2 illustrates how a person using technology to carry out a specific task can be seen as being embedded in a complex system of overlapping and interrelated motivators and constraints.

In most cases, the technology involved will be some form of personal mobile technology, such as a smartphone, tablet, wearable device, or even a laptop. In addition, the physical spaces where the user is located, and moving through, will often contain additional technology options to supplement the user's own personal mobile technologies. This is especially true in libraries and on university campuses, which often provide public workstations and touch screen displays for people to use.

At the center of this model of the mobile context, however, is the user and the task that he or she is trying to accomplish. The user is typically performing this task while engaged in another activity in a physical environment that is continually changing and providing stimuli. The user's attention is not likely to remain fully on the task, but instead will shift from one part of the context to another, as demanded by the task or activity or made necessary because of other interruptions. It is this variable and ever-changing nature of the mobile context that perhaps differentiates it the most from the desktop context (Tidal, 2005).

**FIGURE 1.2**
The mobile context. Adapted from "Design Sketch: The Context of Mobile Interaction," by J. Braiterman and N. Savio, 2007, *International Journal of Mobile Marketing, 2(1)*, p. 67.

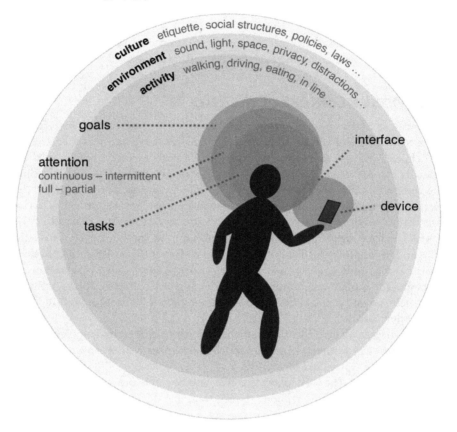

The mobile context also recognizes that users are individuals and that they will differ in meaningful ways. The users' previous experience and familiarity with the library, with the library's systems, with technology, and specifically mobile technology, will all impact both the technology choices that they make and how effective they are in using the technology. Their own cognitive abilities, emotional state, and motivation to complete the task also play a role in the choices they make and their ability to use the technology effectively. The users will also be influenced by cultural norms of what is considered appropriate behavior, selecting and using technology in a way that suits their behavioral response that larger cultural context.

This model of the mobile context demonstrates how clearly inadequate and ill-suited the traditional, contextless user-computer model is for informing our work with mobile technologies. The desktop context described earlier can in fact be seen as a specific instance of the mobile context model, one where the choice of technology, location, activity, and environment are predetermined. In order to use computing technology to carry out the task, the person has to go to where the desktop computer is located, sit still at that location, and focus his or her attention on performing the computing task at hand. However, this is an artificial construct even for desktop users today, who are not necessarily sitting quietly and are rarely in an environment that is free of distraction. Even within the virtual "nonspace" of the human-computer interface, people are likely to be trying to do several things at once, to be subjected to various forms of virtual distraction (e-mail messages, reminders, notifications, etc.), their attention continually shifting between all of these elements. People working at a laptop or desktop computer may also opt to use their mobile devices as they are perceived as being more convenient or a more "natural method" for carrying out a specific task (Heimonen, 2009). Walsh (2012b) described this use of a mobile device to supplement a laptop or desktop as "dual screening" (p. 14). He also observed, at least within an observational group of more advanced mobile technology users, the practice of moving content between devices so that the content could be utilized on the user's preferred device (Walsh, 2012a).

Mobile technologies are no longer a poor alternative that people use when they do not have access to a desktop computer. From the user's perspective, depending on the task, the mobile device is equal to, or even superior to, the desktop computer. The mobile context is the perspective that allows us to understand and design for that reality.

## *Tasks and Micro-Tasks in the Mobile Context*

Mobile devices give people the choice to decide how they want to use them and integrate them into their lives. Kim (2013, pp. 10–12) notes the wide range of behaviors that constitute contemporary mobile device use. Sometimes, people using mobile devices are in a rush and have only a short period of time in which to complete their tasks. Other times they are willing to spend extensive amounts of time viewing content, communicating, or playing games. People may turn to their mobile devices for distraction when they are bored, or they will rely on them when they are fully engaged in some activity. Although there was a time when the limited capabilities of mobile devices meant that they could be used to carry out only simple tasks, the functionality of today's devices matches that of many desktop computers. People are using their devices to run more complex applications and in some cases have turned to their mobile phone or tablet as their only personal computing device.

In order to fully understand the mobile context, libraries need to seek a deeper appreciation for what motivates people's use of mobile technologies. For example, Silva and Firth (2012) describe how smartphones, tablets, and other mobile devices that we have here at the start of the twenty-first century provide people with a wide range of ways to filter, perceive, experience, and otherwise mediate their interactions with the space around them. Location-aware devices make possible new ways for people to access information related to their current context and to connect with objects and people in the space around them. The emergence of smart devices, appliances, and objects that are said to make up the *Internet of Things* (Kopetz, 2007) has moved us further along toward a future of ubiquitous computing, and already we are seeing how our personal mobile devices integrate with this world (Islam & Want, 2014). The idea of mobile technology as a means of mediating our experiences with the world around us provides a useful, broader perspective on why and how people seek to use their mobile devices and serves to ground the more specific observations we make about mobile device use.

The task a user is trying to complete can be thought of as consisting of several micro-tasks. Micro-tasks are a series of smaller tasks of limited scope that are each part of a larger workflow that is carried out to accomplish a broader task or goal (Kim, 2013, p. 11). For example, a student wanting to take a book out of the library might carry out following micro-tasks:

1. Get the title of the book.
2. Find out if the library has the book.
3. Reserve the book.
4. Receive notification that the book is ready to pick up.
5. Find out when the library is open.
6. Find out how to get to the library from a friend's place.
7. Find out when the friend is available to go with him or her to the library.
8. Take note of when they will be visiting the library.

This planning work done, the person would then need to travel to the library, meeting the friend somewhere along the way, locate the book in the library, and then check the book out. Each of the activities could possibly be broken into smaller micro-tasks, all of which would occur over an extended period of time, interrupted by whatever other activities that the person engages in as he or she goes about the day.

The above example is meant to not only demonstrate the complexity of analyzing tasks in the mobile context, but to also emphasize the importance of doing so from a user's perspective. A technology-centric approach will describe the task in terms of the actions that involve the technology. An organizational-centric approach will describe tasks in terms of the steps, requirements, and constraints of the organizational process they are related to. Only a user-centric perspective would, for example, identify elements like the user's desire to involve a friend as he or she seeks to complete the task. These types of micro-tasks, traditionally considered external and outside the scope of consideration, are an essential part of the mobile context and can no longer be overlooked.

It is possible to imagine a number of ways that each of these micro-tasks could be carried out, many of which could involve the use of some form of mobile technology. The challenge for us is that for each micro-task, people will choose which available technology to use based on their preferences and their assessment of how useful or well-suited a device or application is for a given task (Walsh, 2012b). These choices are likely to vary as the context changes and will be different for each person. Some people are comfortable with computing on the go, some prefer to at least be stationary, while others prefer to wait until they are sitting in a more traditional setting even to use their mobile phone, tablet, or laptop. Some prefer or have to use one device for everything, while others are able, and prefer, to use different devices for different tasks (for example: a phone for communications and looking things up, a tablet for reading, a laptop for writing).

The widespread adoption of mobile technologies has changed people's expectations about how they can use technology to communicate, accomplish tasks, and interact with the world around them. Libraries that ignore this reality and continue to implement technology based on the traditional, oversimplified model of desktop computing are likely to find themselves increasingly unable to meet the needs of their users. Libraries that make a commitment to explore and understand the mobile context of their users can leverage that understanding to improve not only their mobile and desktop technology deployments but also their user experience across all aspects of the library.

## Exploring the Mobile Context of Your Users

Although acknowledging the inadequacy of the traditional desktop model to describe how people use technology is an important step, the mobile context only identifies the various elements and relationships that need to be understood in order to successfully conceive and implement mobile technologies. Relying on professional expertise or findings from the literature can inform our work, but only in a general sense. In order to develop a detailed and accurate understanding of the mobile context, libraries need to explore and immerse themselves in the mobile context of their users.

Libraries can begin by undertaking exploratory studies to better understand the mobile context of their users, to understand how they go about learning, teaching, and carrying out research, and specifically how they make use of mobile technologies to support these activities. These exploratory studies would allow libraries to understand their users' needs and behavior in a particular context to see if there are opportunities to improve service using mobile technologies (or other solutions). Bentley and Barrett (2012, pp. 20–21) provide an example of such a process, describing a staged process used with design teams at Motorola and MIT that begins with a broad exploration of an area of interest to identify issues and opportunities to deliver new or improved products. From these initial explorations, they develop specific research questions that guide a second round of more specific observations, focus groups, and interviews. The understanding gained through this process is then used to inform the design of new products and services.

In many cases, libraries may consider themselves to already be past the exploratory stage. Having identified what they perceive as a need or opportunity to improve library service using mobile technology, these libraries will have a good idea of who and more importantly which behaviors they need to observe. Libraries may want to be cautious, however, and not focus too closely on their original ideas, but instead make their observations open enough to allow for some exploration of the problem space, if only to validate their original assumptions about the nature of the problem, as well as the potential for using mobile technology to address it.

The people and situations that we choose to study will depend on the motivation for undertaking the study. Observations should include all behaviors and aspects of the context that appear to be meaningful to the participants or that influence their behaviors in some way. For example, if you are specifically interested in improving services for faculty, you'll need to observe faculty members in a number of settings. If you are interested in supporting learning in the classroom, your study would include students, but also faculty and teaching assistants, with the observations being made in the classroom during (and potentially before and after) class.

Libraries traditionally turn to surveys, focus groups, and interviews as ways of better understanding the needs and preferences of users. Unfortunately, none of these methods are appropriate for exploring how people use mobile technology. You also cannot get an accurate picture of how people use mobile technologies in an artificial, controlled laboratory-like setting (Zhang & Adipat, 2005). If we are to develop an accurate understanding of the mobile context of our user communities, we need to observe the mobile context directly. Bentley and Barrett (2012, p. 42–43) recommend using task analysis as a framework to explore the mobile context. This can be done by asking participants to carry out the relevant tasks in the actual context of interest. Researchers can observe behavior directly, while asking participants to speak aloud while they carry out the task. Doing this gives the researcher access to what people are thinking. Partially structured interviews can be used afterward to seek further explanation and insights into observations made during the task analysis.

Even without the participation of users, librarians and project team members can benefit from situating themselves in the same contexts as their users as they try to identify and develop solutions. For example, brainstorming sessions held in real-world contexts can provide teams with an appreciation of the opportunities and limitations within which their applications will need to work (Hinman, 2012, p. 56). Librarians are then likely to have a more realistic sense of how willing their users are likely to be engage with any given solution. It is also harder to ignore the constraints imposed by the context because librarians are actually experiencing them as opposed to having to imagine them.

It should be noted that many libraries will likely find it challenging to dedicate the time and resources required to carry out extensive field observations. However, any amount of time spent observing or directly experiencing the real-world conditions in which mobile technologies are used is likely to be beneficial to the design team. Even a one-hour observation session can provide insights and shift people's thinking enough to help them improve their understanding of the mobile context and therefore make better decisions about what to develop and how it should be designed (Bentley and Barrett, 2012).

# Informing Your Mobile Technology Strategy

Libraries that want to achieve their goal of serving their user communities can no longer assume that people will adapt their preferences and needs to the services the library chooses to offer. Instead, libraries need to be ready to adapt their own services and operational models to meet the needs of their users (Mello, 2002). The mobile context provides libraries with a perspective that is focused on the users, the tasks they are trying to achieve, and the complex environment they are living in. From this perspective, libraries can gain a better appreciation for how their community uses the library and the ways in which the services and resources offered by the library fall short of meeting user needs. They can also better conceptualize changes that are likely to improve the user's experience.

There are several ways that a library can provide support for their services and resources on mobile devices. The decision on the most appropriate technology strategy is often based on the library's development resources and capabilities. However, an accurate assessment of the library's understanding of their users' mobile context can play an equally important role in choosing how to proceed. Kim (2013, p. 12) summarizes the trade-offs that libraries often consider when deciding on the technological approach that they are going to use for supporting mobile devices. Many academic libraries possess, or can easily acquire, the technical skills necessary to implement responsive web applications in a reasonable time frame. However, web apps run in a browser and have limited access to the full capabilities of the mobile device. Native applications can take full advantage of the mobile technologies and often provide a superior user experience. However, providing this experience requires the library to commit to a significant and ongoing investment of resources. Native apps require specialized knowledge and are harder to develop. They are also platform-specific, meaning that libraries need to implement and maintain separate versions of their apps for each platform they need to target. As a result, few libraries are able to develop custom native applications, choosing to implement responsive, mobile-ready versions of their websites instead.

More importantly, it takes more than technical proficiency and capabilities to successfully implement a mobile solution. In order to be effective, libraries should take care to deploy technology solutions that are in line with their understanding of the mobile contexts of the community they serve. Figure 1.3 shows the level of understanding of the mobile context required to be able to effectively deploy the various types of mobile solutions.

### FIGURE 1.3
Understanding of mobile context required for mobile solutions.

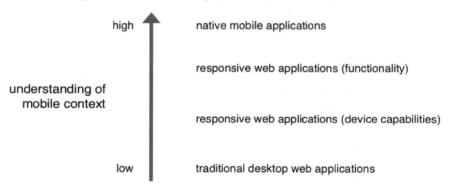

Libraries with little or no understanding of their users' mobile context should limit themselves to deploying traditional desktop applications and websites. While users will still be able to access these websites on their mobile devices, they will be hard for people to use (Heimonen, 2009). A basic understanding of the mobile context of their users, and specifically an understanding of the mobile devices they use, can allow libraries to create responsive web applications that are tailored to the display and input capabilities and limitations of the devices.

As the library develops a more comprehensive understanding of the mobile context, it can move beyond "retrofitting old experiences into a new mobile medium" (Hinman, 2012, p. 43). It can use responsive technologies to customize its applications, adding, removing, and modifying functionality to suit the needs of the context. Libraries that go further, developing and maintaining a deep understanding of the user context, can invest in developing and deploying native apps that take full advantage of the capabilities of the mobile device. Far from being a straight port of the library's desktop applications, these native apps can be tailored to the specific needs and realities of the patron's mobile context.

Libraries that lack the technical expertise or capacity necessary to design and build mobile applications themselves can opt to license third-party mobile applications. However, this does not alleviate the need for the library to invest in exploring and understanding the mobile context of its users. Understanding the mobile context will allow libraries to identify what applications are needed and how these applications need to deliver their functionality if they are to be used by students and faculty. Libraries will also be able to leverage their experience in observing, exploring, and understanding the mobile context to work with users to assess the various third-party applications as part of the selection process. By grounding the selection of third-party applications in the mobile context, libraries can ensure that the needs of their users are the primary consideration when establishing selection criteria and that these requirements will carry sufficient weight to balance other factors such as technical, organizational, and financial requirements in the final assessment.

## Conclusion

Despite the rapid adoption of smartphone technology, we are still in the early days of mobile computing. At the time of writing, it has been less than ten years since the iPhone was released (Allison, 2007). Recently, we have seen the emergence of both wearable devices as well as the proliferation of embedded technologies that over time will provide people with even more computing options in any given context. We are still learning, both as individuals and as organizations, how to best incorporate this technology into our work and our lives. Our approaches, techniques, and best practices for designing, developing, and deploying software applications are still very much rooted in the traditional desktop model. The mobile context provides librarians with a user-centric perspective from which to better understand their relationships and interactions with the communities they serve, preparing them to better meet the challenges and opportunities presented by technological advances the future is sure to bring.

## References

Allison, K. (2007, January 9). FT.com site: Apple rolls out much-anticipated iPhone. *FT.com*. Retrieved from http://search.proquest.com/docview/229017840?accountid=12339.

Bentley, F., & Barrett, E. (2012). *Building mobile experiences.* Cambridge, Mass.: The MIT Press.

Braiterman, J., & Savio, N. (2007). Design Sketch: The Context of Mobile Interaction. *International Journal of Mobile Marketing*, 2(1), 66–68.

Dey, A. K., Abowd, G. D., & Salber, D. (2001). A Conceptual Framework and a Toolkit for Supporting the Rapid Prototyping of Context-Aware Applications. *Human-Computer Interaction*, 16(2–4), 97–166.

Greenfield, A. (2006). *Everyware: The dawning age of ubiquitous computing.* Berkeley, CA: New Riders.

Harrison, R., Flood, D., & Duce, D. (2013). Usability of mobile applications: Literature review and rationale for a new usability model. *Journal of Interaction Science*, 1(1), 1.

Heimonen, T. (2009). Information needs and practices of active mobile Internet users. Paper presented at the *Mobility '09: 6th International Conference on Mobile Technology, Application & Systems*. Article 50 (pp. 1–8). Retrieved June 15, 2016 from ACM Digital Library.

Hinman, R. (2012). *The mobile frontier: A guide for designing mobile experiences.* Brooklyn, N.Y.: Rosenfeld Media.

Islam N., & Want R. (2014). Smartphones: Past, present, and future. *IEEE Pervasive Computing*, 13(4), 89–92.

Kim, Bohyun. (2013). The Present and Future of the Library Mobile Experience. *Library Technology Reports*, 49(6), 15–28.

Kopetz, H. (2011). Internet of things. In *Real-time systems: Design principles for distributed embedded applications* (pp. 307–323). New York: Springer.

Mello, S. (2002). *Customer-centric product definition: The key to great product development.* New York: AMACOM.

Silva, A. de S. e., & Frith, J. (2012). *Mobile interfaces in public spaces: Locational privacy, control, and urban sociability.* New York: Routledge.

Tidal, J. (2015). *Usability and the mobile web: A LITA guide.* Chicago: ALA Tech Source.

Walsh, A. (2012a). Mobile information literacy: A preliminary outline of information behaviour in a mobile environment. *Journal of Information Literacy*, 6(2), 56–69.

Walsh, A. (2012b). *Using mobile technology to deliver library services: A handbook.* Lanham, Md.: Scarecrow Press.

Zhang, D., & Adipat, B. (2005). Challenges, Methodologies, and Issues in the Usability Testing of Mobile Applications. *International Journal of Human-Computer Interaction*, 18(3), 293–30.

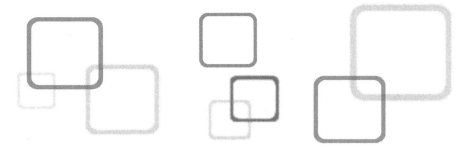

CHAPTER 2*

# The Development of an Academic Library's Mobile Website

*Junior Tidal*

## Introduction

Mobile devices have become more ubiquitous among academic library users. It's now common to see through analytics that smartphones, tablet computers, e-readers, and even portable gaming consoles are connecting to online library services. This chapter is a case study of how a small academic library supported its users through the creation of a mobile-optimized library website. It documents the chronological changes from the website's humble beginnings on a shared Windows IIS server to its current configuration on a Linux-based cloud server. Throughout its existence, the website was developed with adaptability in mind, and flexibility, in order to respond to unpredictable changes in information technology. This adaptability includes not only the changing landscape with regard to standards for mobile website development, but also changes in user preferences over time with regard to devices and website architecture.

## Literature Review

Content management systems (CMSs) have changed the way that libraries have crafted their online identities. These systems centralize webpage content creation

---

* This work is licensed under a Creative Commons Attribution- NonCommercial-NoDerivatives 4.0 License, CC BY-NC-ND (https://creativecommons.org/licenses/by-nc-nd/4.0/).

and allow for sites to be better organized. Libraries have employed CMSs to facilitate the presentation of complex websites in a user-friendly fashion (Black, 2011). A CMS can also put content production in the hands of all librarians and not just a single "gatekeeper," flattening technological hierarchies of control. This distributed system of creation was important for the Darian Public Library's website, where Drupal was selected to support its active internal blogging community (Sheehan, 2009). Another library with goals similar to our own, the library at the University of California, Santa Cruz, adopted Drupal because it could bring consistency to older webpages and be used to remove defunct and redundant webpages (Hubble, Murphy, & Perry, 2011).

Libraries have crafted mobile library websites to reach their users. Librarians and library developers who have studied users so that they can meet their organization's specific needs explain that understanding your user base will help in designing a mobile site that will be "heavily trafficked by your users"(Bridges, Rempel & Griggs, 2010, p. 318). Needs assessments have also been conducted through surveys to gather users' preferences (Dresselhauls & Shrode, 2012; Cummings et. al, 2010). Preliminary research on device usage has been conducted to justify a mobile library site (Wilson & McCarthy, 2010).

In discussing mobile web development for libraries, it is difficult not to touch upon the adoption of responsive design. Glassman & Shen said it best, stating that the "mobile web has been replaced by the responsive web—one site fits all" (2014, p. 89). Responsive design is a practice where websites are coded to conform to the screen size of the device being used to view it. The page adapts (or responds) to a layout specific to these dimensions. The flexibility of responsive design makes updating a library site easier and makes the site functional regardless of the device it is viewed on (Kim, 2013).

## Background

The Ursula C. Schwerin Library serves the research needs of the New York City College of Technology, City University of New York (CUNY), located in downtown Brooklyn, New York. Locally known as City Tech, the college supports over 17,000 students in associate and baccalaureate programs across a wide variety of programs. Students have access to over twenty CUNY campus libraries throughout the five boroughs of New York City, in an arrangement similar to that of a consortium resource-sharing library system. The library's integrated library system, monographs, and some electronic resources are centralized.

The student population of City Tech is quite diverse. Almost half of the student population was born outside of the United States. The majority of the student population identify as black, Hispanic, or Asian. Over 60 percent of students speak a language other than English at home. Most freshmen receive need-based financial aid. The college can be characterized as a commuter school with many students working full- or part-time jobs, with a significant number of students being enrolled in continuing education programs.

# History of the City Tech Library Mobile Site

To understand the development of the library's mobile site, it is best to first examine the creation of the main library website. The library's homepage (https://library.citytech.cuny.edu) was first hosted on a Windows IIS server in 2002. The site consisted of several static HTML pages that were manually edited using Microsoft FrontPage. The maintenance of the library website reflected its numerous curators, as the work was shared between the technical services librarian and the multimedia librarian. There was a lack of uniformity not only in the aesthetics of the site, but in its information architecture as well. Even though the site had its own URL domain, pages were not properly grouped in appropriate directories. Numerous pages were orphaned, contained broken links, or were simply outdated.

Workflows for creating and amending content were cumbersome in this early iteration of the library website. Librarians had no direct access to the site through a browser. Instead, content was e-mailed to the web services librarian, marked up in either FrontPage or Adobe Dreamweaver, and then uploaded via FTP to the college's server. As the amount of content grew on the library website and services such as electronic resources began to proliferate, it was apparent that the web server was in dire need of upgrading. There were more demands for space and for processing power to serve webpages.

The library web server, hosted by the campus Computing and Information Systems (CIS) department, was shared with several other City Tech academic departments. Due to this configuration, a number of security restrictions were implemented by CIS. Many types of software were prohibited from being installed on the server, as database software and web scripting languages were seen as possibly disruptive to other departmental websites. This prevented the installation of various programs, such as CMSs, analytics tools, and web form processing. Shell access outside of the college's IP range was unavailable, making updating content or repairing server problems from off campus impossible.

## *Acquiring a Server for the Library*

In 2007, after a written proposal from the Web Services and Multimedia Librarian, and the Chief Librarian, the provost allowed the library to procure its own web server. This Dell server ran on a Red Hat Enterprise Linux (RHEL), Apache,

MySQL database, and PHP configuration also known as a LAMP setup. This is a popular web server configuration, utilizing open-source technologies. Red Hat is a specific distribution of the Linux operating system. Apache is widely used open-source software that delivers websites. MySQL (server query language) is a relational database language, and PHP (Hypertext Preprocessor, a recursive "backronym"), a web processing language.

This new server configuration lifted the restrictions that had impeded the previous shared server. With access to PHP and MySQL, the library website could now implement blogs, CMSs, web analytics, and server log reports. A PHP script was written to display uniform headers and footers across the site. Navigation menus and library information were consistently displayed on every page of the site; lack of consistency was a problem that had plagued previous iterations of the library website. More importantly, the implementation of this new server had opened an opportunity to create the library's first mobile website.

## *Mobile Site 1.0*

Based on information gathered from analytics data, we found that mobile device connections to the library website were increasing. It's important to note that these visits were occurring prior to the release of the first generation iPhone in 2007. Some of the devices connecting to the site were cellular phones with web capabilities. Surprisingly, some of these devices were portable gaming consoles that had Wi-Fi and web browser capabilities. Concurrent with this activity, electronic resource vendors were beginning to release light mobile versions of their products. These combined factors spurred the library's first attempt at developing a mobile webpage.

The first mobile page for the City Tech Library consisted of static HTML pages. It was developed in a text editor and contained neither CSS declarations nor images. This first page was very basic and contained only three links: to the library's shared CUNY-wide catalog, to a page containing electronic resources, and to a page containing library contact information and hours. The page displaying electronic resources was limited to those resources that were mobile friendly. This first mobile site was hosted at the URL http://library.citytech.cuny.edu/mobile. A PHP script placed on the homepage of the library website redirected users to this page based on the user's browser agent type. User testing was not conducted in the creation of this first mobile site. This first mobile page on the library website was short-lived for a number of reasons. First, the PHP script to redirect users to the mobile site didn't always work. The script functioned by detecting what kind of browser a user was using to visit the site. Unfortunately, some users logging in with a cell phone browser were not redirected to the mobile site. Second, this agent detection script required constant maintenance. New cell phones that were web-enabled were being released, as well as a plethora of other various mobile devices and browsers, all of which needed to be added to the redirection script manually. At the time of this first mobile site's existence, no statistics were taken as to how often the page was visited. This was problematic as there was no data to support the continuing existence of the page.

## *Drupal*

In 2010, the library migrated away from its custom PHP-scripted setup and to the Drupal 6 (D6) CMS. Drupal (http://www.drupal.org) is a modular, open-source CMS that utilizes MySQL and PHP to manage web content. It is modular in the sense that there are numerous modules, similar to WordPress plugins, that can enhance the functionality of D6. A large community of developers is creating these modules, including a few who have ties to academic libraries. Since these components are open-source, a library site can be fully customized to meet specific user needs. Blog and calendar feeds can be aggregated from other sources and displayed on the library website. The presentation of content is customized through CSS scripts and PHP code. Code of an existing module can be modified to render an appropriate display of that aggregated content. For example, a series of library events from a Google calendar can be aggregated, parsed, and reproduced in a table form on the site. Finally, one of the more powerful aspects of D6 was the ability to modify content within the browser. This is probably the most practical reason to adopt a CMS. It may seem ubiquitous among content management servers to provide this today, but at the time, this feature was very useful for librarians with different levels of technology experience. Users could simply log into the library website via the browser and update content as needed.

Using a CMS for editing was very different from updating HTML pages or PHP scripts, which previously required a connection into the web server through a shell and then editing the scripts through a command-line text editor. The library's electronic resources had been kept in a flat-file database. This flat-file database was simply a text file that contained information about a single electronic resource per line, including the URL, the name of the resource, the date added, and a description. The electronic resources librarian would have to manually enter each resource line by line to update it. This process was prone to numerous errors. Existing resources required a search and find command to make any changes to the resource's URL, name, or description. Through D6, the workflow to update these databases was significantly improved. Now, librarians could log into the site through their mobile device or workstation and make changes faster and more efficiently through the web browser. Electronic resources were stored within Drupal's custom Content Creation Kit module, which was much more user-friendly than editing flat files.

At the time of this Drupal migration, analytics tools were used to gain insight into our users and their behavior. Log server files, Google Analytics, and its open-source alternative Piwik were all used to allow us to better understand visitors to the library's site. These tools were implemented using Drupal modules designed for this purpose. This analytic data has been used to drive the development of the mobile website (Tidal, 2015). For instance, analytics can show what devices are connecting to the site, how fast the connection is, the network provider, and a masked IP address that can provide geographic locations. The site can then be tailored to these factors by approaching site content from a "mobile first" perspective. "Mobile first" is a practice where content is developed from the perspective of a small screen. The design, layout, and content are constrained and, as screens get larger, are amended with more features. This also impacts bandwidth, as constraining file size to improve website performance on slower cellular broadband networks can be factored into the creation of the site.

Although mobile device visits to the library website make up a small percentage of overall traffic, their number has increased steadily since 2011. From 2011 to 2016, mobile traffic increased from 1 percent to 7 percent of overall traffic. This is reflective of overall trends of cell phone ownership. The Pew Internet Research and Life Project reports an increased number of cell phone owners, surpassing that of desktop and laptop workstation owners (Smith, 2015). Many of these owners also use cell phones as their primary access point to get online (Smith, 2015). This trend required the City Tech library to respond by creating another iteration of the library mobile site.

## *Mobile Site 2.0*

The second iteration of the library mobile site was developed in 2012. Hosted on the same server as the main library website, a virtual server was configured in Apache serving this separate mobile site. The mobile and desktop sites' file structure were both contained within Apache's default /var/www/html directory. Yet the mobile site was contained in a subdirectory aptly labeled *mobile*, whereas the desktop site was contained in a subdirectory called *libSite*. This set the foundation for two separate Drupal installations.

This version of the mobile site also used a PHP redirection script provided through a D6 module. It was similar to the script used in the first iteration of a site: if a user visited the library website using a mobile-enabled device, it would redirect the user from http://library.citytech.cuny.edu to http://m.library.citytech.cuny.edu. The design of the site was also user-centered from the ground up. For instance, because we simply added the letter *m* to the virtual domain of the mobile site, users would need to type less within the browser's address bar to get to the mobile site. This made the site more usable compared to a site that simply amends the mobile site's location as a subdirectory, such as http://library.citytech.cuny.edu/mobile. It also reinforced the fact that the mobile site was separate from the desktop one.

Even though the main library website was utilizing D6, Drupal 7 (D7) was chosen to manage the mobile site. D7 was chosen over D6 because it supported a number of modules and themes that specifically optimize webpages for mobile devices. Coincidentally, a number of D6 modules were unfortunately unavailable for D7. This had little bearing, however, on the mobile site since it was intended to be minimalist in design. It was preferred that the mobile site have a small footprint to accommodate mobile users connecting through cell networks or spotty Wi-Fi connections.

One D7 module that was used to accommodate this smaller footprint is the popular jQuery Update (https://www.drupal.org/project/jquery_update). jQuery is a JavaScript-based library used to easily implement JavaScript in a webpage. It is useful in navigating the components of a webpage as well as enhancing a page's functionality. It is cross-platform-compatible so that any device can process its scripts. These scripts allow a device to display animation, process event handling, and navigate a webpage's source document object model (DOM). The scripts are especially useful in providing feedback to the end user viewing the page. Unfortunately, there is no support for later versions of jQuery for sites using D6. This is not the case for D7 installations. This was very important because of the intended use of jQuery Mobile, a separate JavaScript library derived from jQuery. jQuery Mobile, as its name implies, supports mobile devices. It gives developers the tool kit to design touch-based interfaces, responsive websites, and applications.

A D6 module was installed on the desktop website to detect mobile devices. This module redirected users to the mobile site if they were detected using a mobile device. It was much more efficient than the in-house-created PHP script in the first mobile site iteration. A link to the library's desktop site was also present in the footer of each mobile page. Members of the library's website committee felt that it was necessary to have this link in case there were online transactions or tasks that could not be accomplished through the mobile instance. Although the D6 desktop site was not mobile-optimized, users were still able to access content not found on the mobile site.

The content of the D7 site (figure 2.1) was based on the most popular pages visited on the main library website. This data was collected through web analytics tools Piwik and Google Analytics. The popularity of these pages dictated the order in which links appeared on the page. The webpage's header contained the library's logo with the current day's operating hours underneath. Since the "Hours" was the most visited page according to analytics of mobile devices, it was given the greatest priority in the visual hierarchy of the page. However, analytics tells only part of the story. In order to see how effective changes were based on analytics data, we later conducted a usability study to confirm whether or not our changes aligned with user preferences. Log files and analytics data alone don't necessarily justify design decisions.

**FIGURE 2.1**
Screenshot of the D7 mobile website.

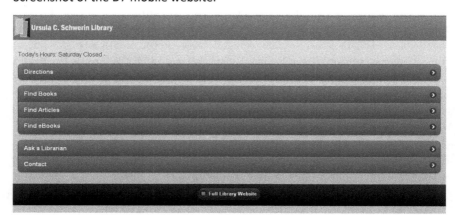

Links on the new mobile site were separated into groups of large buttons. The first version of this D7 mobile site contained a link to a page with library directions. A second set of buttons immediately below the first included links to a catalog search, a page containing mobile-optimized electronic resources, and a linked button leading to electronic books. The last button group contained links for reference desk contacts and a contact button for the library's circulation services.

A PHP script was developed to manage electronic resources between the mobile and desktop versions of the site. To alleviate the workload of managing two sites, this automated script would replicate the electronic resources found on the desktop site, parse them, and display the results on the mobile version of the page. Only resources that contained a mobile version would be replicated on the site. This was especially useful at the time when e-books were increasing in popularity (Tidal, 2012).

The layout of the mobile site was designed to be responsive to screen size, since jQuery Mobile is inherently responsive. Large buttons that were easy to tap were employed for smaller screen sizes and resolutions. This approach also supported mobile device users using touch interface because if links are too small, users could possibly have problems accessing the links with their fingertips. There was also a lack of images to create a lightweight site that focused on loading in a speedy and efficient manner.

## *Evaluating the Mobile Site*

After the mobile site was launched, a usability test was conducted during the 2013–2014 academic school year (see Tidal, 2014). The objective was to ascertain if the mobile site could support students' research needs. In order to entice volunteers, users were given campus bookstore gift cards as incentives to participate in the study.

A cognitive walkthrough was used as the evaluation instrument for the mobile site. This was moderated by a testing proctor and digitally recorded. Participants were asked to complete a series of task scenarios derived from usability goals. For instance, one goal could be users being able to find a book on the library website. A task scenario would be written to place the user within a realistic context to achieve this goal. For example, a task could be "Imagine you are in an English class and you need to complete a research paper on the book *The Hitchhiker's Guide to the Galaxy*. Using the library website, find the call number of this book." Here, the scenario gives the user some context for completing the usability goal and is a call to action to complete the task using the site. Users are asked to think aloud as they attempt to complete the goal while their responses are recorded. This process provides proctors a glimpse into the thinking of the users as they interact with the website.

Participants in the usability test consisted of a variety of students: ten students, five for the first round of usability testing and five for the second round of testing. This approach was modeled after the common practice of using a limited number of participants (Nielsen, 2000). The basic idea is that having more than five users test the site won't reveal any new usability problems, but will simply confirm the existence of issues that have already been identified by previous test participants. However, this assumption has been contested (Faulkner, 2003; Bevan et al., 2003). The sample included traditional two-year and four-year students, as well as nontraditional, or continuing education, students. Prior to the study, a screening survey was given to participants, asking which devices they have used. This allowed them to be paired with tablets with which they were familiar so that students using devices with which they were unfamiliar would not taint the test.

After the first round of testing, the mobile site was restructured based on the data collected from participants. Notable changes included the addition of a search box that takes students to the catalog instead of to a separate page accessible through a hyperlink. Numerous participants found it more useful to have this search box, rather than a link, on the mobile site homepage directly. Participants also revealed that students preferred having a button to take them directly to their library account to view loaned items and renewals. This option was not available on the site, requiring participants to log in through the catalog's search screen. Electronic resources were also organized by device type so that users of Apple or Android products could select which page suited them. Links on the homepage were also rearranged. The Directions page link was moved to the bottom of the page, as it was deemed the least important page by participants.

Even though the site was restructured, the overall usability of the mobile site was well received during usability testing. Participants in the first and second round completed tasks at high rates of success. The thinking aloud protocol (TAP) metric gave further insight into users' reactions of the site. TAP, also known as the think aloud method, is simply where users speak their thoughts aloud while performing various tasks. Lewis and Rieman, who were the first to employ the method, stated that when testing a design, users' "comments are a rich lode of information (1993, p. 83)." Users express what they are doing, thinking, and feeling, which provides feedback to observers. This process not only helps identify usability issues, but helps to pinpoint potential obstacles as well.

The TAP protocol for this study provided librarians with information on how to improve the site (figure 2.2). Users responded that the site was easy to use and that the site was pleasing to the eye. Conversely, they also noted that the site failed in their expectation of its having spelling correction when using the site's online catalog.

**FIGURE 2.2**
Screenshot of the D7 mobile website after usability testing. Note the changes in comparison to figure 2.1.

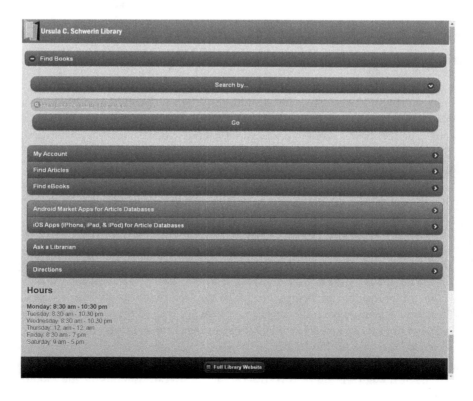

## The Mobile Site Gets Attacked

During the latter half of 2014, the mobile website went down following a malicious attack on the library's Drupal server. A hacker vandalized the mobile site, and the homepage was replaced with online taunts. The library's mobile site was not the only one affected. This security vulnerability compromised an estimated 12 million sites running D7 ("Millions", 2014). As per university protocols, the server hosting the site was disconnected and the site was taken offline. This resulted in the whole library domain going dark as the library's multiple sites were hosted on a single machine. However, disabling the mobile site rectified this problem. This attack ended the last iteration of the mobile site before we moved to a responsive design solution.

## Transitioning to a Responsive-Design Website

Even though mobile users make up a small portion of overall visitors, it was important to support that minority group of users. In response to the recent hack of our mobile site and the demand for mobile support, the possibility of implementing a responsive-design website was explored. At the time, numerous other academic libraries were adopting responsive-design websites as a way to cater to both mobile and desktop users. This notion of providing a more customized experience for users on different devices was extremely appealing to the City Tech Library team. Despite homebrewed PHP scripts that parsed content, it was very difficult to maintain and update content for both the desktop and mobile sites. This amalgamation helps to prevent posting inconsistent content. Through the consolidation of the desktop and mobile sites into a single responsive site, librarians can actively add more content without switching between the two.

During the 2014–2015 academic year, a prototype of a redesigned website was developed. This version employed responsive-design techniques. Responsive design is a technique where a website presents a customized experience for the user, tailored to the device used to visit the site. A website will be displayed with the same content and a similar layout on a desktop, smartphone, or tablet computer. This is accomplished by CSS style sheets, rules, and declarations, which are triggered by specific device widths. For instance, a device width of 480 pixels, common for smartphones, activates CSS rules that cause the site to "respond" to that width. Based on the detected dimensions of the screen, the site will shift its layout and conform to the display of the device. Altering the browser window size on a desktop workstation will also cause the site to conform to the new window width.

Responsive design alleviates the problem of device detection. Detecting what type of device is being used to access the site through agent requests is not perfect. The previous site's detection methods, as previously stated, were problematic. Not only will responsive design overcome such shortcomings, but it will also cause less of a dependence on updating the ever-growing list of devices that can be used to access the mobile Web. Responsive design enforces consistency between the mobile and desktop site, and this encouraged the library's adoption of D7. At this point modules used in the D6 version of the library website were available for D7. This hindered the early adoption of D7. Alternate modules to accommodate the D6 installation were developed and implemented in this new D7 site. Different modules and themes were tested to use responsive design for the prototype site.

The most successful theme is a port of the popular web framework known as Bootstrap (https://getbootstrap.com and https://www.drupal.org/project/bootstrap), which was developed by two Twitter developers, Mark Otto and Jacob Thornton. This free, open-source framework contains several useful features. One feature of note is the framework's contribution to the rapid deployment of a website's front end. Numerous built-in features commonly found on websites, such as form controls, buttons, and navigation components, were readily available out of the box, and so additional code wasn't necessary to create them. Bootstrap is also cross-platform-compatible for a wide range of devices. This is important, because different devices use a wide variety of web browsers. An Android tablet may render a webpage much differently from an iPhone. An Apple MacBook running the Firefox browser may display web content differently from a Linux workstation running the Opera browser. A cross-platform web framework displays webpages as similarly as possible across this wide range of devices. Bootstrap is built with responsive design, so it runs on a wide range of devices of various screen widths.

Adopting responsive design didn't stop at the redesigned website prototype. At the time, the library website housed three WordPress installations. This included the library's blog, an orientation site for incoming freshmen, and a newsletter site updated each semester. All of these sites under the library's domain were updated with responsive-enabled themes. The library staff intranet was initially a MediaWiki installation but was also converted to a mobile-friendly WordPress instance.

Although D7 and its modules made the site more accessible for mobile and desktop users alike, the transition process was not without obstacles. Some features within Bootstrap are incompatible with the Bootstrap theme developed for D7. Numerous related open bug issues can be found on the Drupal Bootstrap project site (https://www.drupal.org/project/bootstrap). As a result, extensive theme modifications were required on the prototype site. To better customize Drupal Bootstrap, a "child" Bootstrap theme was created. A child theme is a subtheme of an existing Drupal theme, known as the "parent" theme, that inherits many of its characteristics and functions. This child theme contains updates to the PHP-based header files that controls the elements first loaded into the browser. These modifications include details such as specific div CSS classes to the header's navigation section, menu modifications, and header spacing. CSS modifications were implemented to change the positioning of default elements, change theme colors, and tweak element margins and paddings.

In addition to these customizations, the library website also migrated from the City Tech CIS-hosted servers to an Amazon Web Services (AWS) instance in the cloud. There were a number of factors influencing our decision to use AWS. First, the server housed by the CIS department needed an upgrade in any case. Other factors included the increased demand for online library services and the library's vision of creating a flexible infrastructure that is not only scalable for future technologies, but reliable as well. However, the main reason to use a cloud-based service was cost. Taking into account the amount of traffic the library website had experienced, along with the growth of content, a subscription to AWS was a fraction of the cost of upgrading the library's web server. There were nevertheless concerns about adopting AWS. CIS wanted to make sure that the library web server was not storing any information that would identify a student, such as Social Security or student identification numbers. The library website collects none of this, so it was given the green light to adopt the AWS platform.

This new web prototype was initially constructed on a Mac Pro workstation. Since OS X is based off of UNIX, the site was migrated to the AWS instance running Amazon's distribution of the Linux operating system. Prior to this, the site was shared with other librarians in the department, who gave feedback on its design and function. The installation on the Mac Pro continued to be useful even after the site was migrated to AWS. It was possible to test modules or site modifications before they were implemented on the AWS production server. With the prototype hosted on AWS and its modifications complete, the next step was to begin usability testing of the responsive-design website. The first round of testing began in the fall of 2014, and the second round concluded at the end of the fall 2015 semester. In between the first and second rounds of usability testing, the prototype went live in the spring of 2015.

At the launch of this new redesign, there were additions to the library website through CUNY-wide initiatives. The discovery tool OneSearch was launched. This is the CUNY brand name of Ex LIBRIS's discovery product Primo. Mobile-optimized, OneSearch uses a responsive design to support various devices. It has supplemented the CUNY Catalog, which is the web OPAC for the university's Aleph system. The catalog itself has also been mobile-optimized by CUNY's Office of Library Services. However, the library website committee decided that OneSearch would be the default search system for library materials since it is more usable than the CUNY Catalog.

The other change to the library website was the adoption of the popular LibGuides CMS. This off-site hosted solution provides the City Tech community with guides to assist them with their research. It replaced the previous MediaWiki installation that housed research guides on the library server. Echoing the redesigned library website, LibGuides 2.0 also utilizes the Bootstrap framework.

This allowed LibGuides to be customized, mirroring the appearance of the library website. It was as easy as copying and pasting Drupal's CSS style sheets into LibGuides. (See figures 2.3, 2.4, and 2.5.)

### FIGURE 2.3
Desktop version of the responsive-design D7 City Tech Library website.

# The Development of an Academic Library's Mobile Website 29

**FIGURE 2.4**
Mobile version of the responsive-design D7 City Tech Library website.

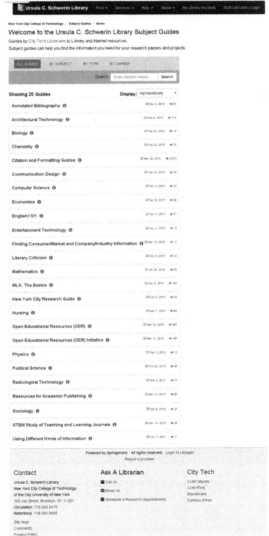

**FIGURE 2.5**
City Tech Library's LibGuides 2.0 mirrors the main site.

There were a number of takeaways following the development history of the City Tech Library's mobile site. These include the need to keep up with emerging technology through active library groups, the need to accept the technological and organizational challenges in developing a website, and the importance of user-centered design. These takeaways could be applicable to any type of library-based technology project. Adapting responsive design to our library website was no small task. A considerable amount of time went into researching responsive design, as well as the Drupal modules that support it. Current practices of other library websites and discussions through various venues assisted in selecting Bootstrap as the framework.

One community that has discussed Bootstrap extensively is code4Lib. code4Lib is a grassroots organization that focuses on coding and librarianship and has a very active journal and e-mail discussion list, as well as national and regional conferences. Some projects mentioned through code4Lib, such as Blacklight (http://projectblacklight.org), utilize Bootstrap. Both the Association of College and Research Libraries (ACRL) and the Library Information Technology Association (LITA) of the American Library Association (ALA) have also discussed the use of Bootstrap on their blogs and e-mail lists. Schofield has written a blog post about how some Bootstrap components are mobile-unfriendly (2014). Keeping up with these professional organizations has helped us to better understand how the library community at large utilizes web frameworks.

One of the biggest challenges of this development project was the attack on the site's Drupal security exploit. The attack was very sudden and unexpected, but luckily did not occur at a busy time for the library. Not only did the attack help us to evaluate the library's security and CMS, it was a catalyst to the redesign. Events such as these should be embraced as an opportunity to improve our services.

Lastly, one of the more important highlights of the mobile development project was incorporating users into the design. It is very easy for us as librarians to take on the role of building what we think is best for our users without actually communicating with them. Using the concepts of user-centered design, where users' input is integrated into each stage of the design process, makes for a more usable website. User input can be the deciding factor in website steering committee discussions, design decisions, and the placement of forms and links. Although I'm not saying that the burden of a website's design should fall squarely on the shoulders of the end users, it is imperative that they be able to use the library website effectively.

## Future Improvements

Future improvements to the library's website will not only run parallel with advances in mobile technology, but also respond to how mobile users interact with sites. There are emerging expectations from the mobile Web, such as hamburger menus, long scrolling, and responsive design, that are now making enough of an impact to make them standards. The hamburger menu or sidebar menu is a popular trend in mobile websites. It is typically a button with three horizontal lines, replacing the navigation menu for devices with smaller widths. Since navigation bars can be long, menu items may not be properly displayed. The hamburger menu alleviates this by compressing the items into a single button. However, this menu system may not be feasible for all mobile projects since it takes another interaction to activate the menu. In addition, the icon itself may not be universally understood, one of the pitfalls of using icons on the Web. Still, the hamburger menu is found in many responsive frameworks and themes.

Mobile devices have changed the way that users interact with websites, specifically, how they react to scrolling long pages. The screen size of a mobile device is considerably smaller than that of a desktop. The notion of "the fold" on a website is no longer a factor. Due the constraints of a smaller viewport and the touch interface of the screen, scrolling is a more intuitive interaction with mobile sites. A long scrolling site also pairs well with simple navigation.

Responsive design is obviously one of the more important trends in mobile web development. Frameworks such as Bootstrap, HTML5 Boilerplate, and the recently developed WC3 CSS framework all employ responsive design. Frameworks make implementation of web applications rapidly deployable. OCLC has been utilizing a custom framework, known as CoreUI, to adjust its interface depending on the specific viewport size (Ganci & McCullough, 2015). Other current trends include content-delivery networks (CDNs), which reduce the file size of CSS and JavaScript. CDNs serve JavaScript and CSS files of frameworks so they do not need to be locally stored. This saves on maintenance and tightens version control. CDNs optimizes file size and bandwidth, a key component in making mobile sites less complex (Kim, 2013). CDNs will probably become more popular with distributed web frameworks. Responsive design in this sense will be more ubiquitous and will require even less coding knowledge to implement.

Libraries can currently adopt responsive design for databases and vendors that do not offer it natively. Reidsma describes customizing HTML and CSS for such cases and integrating them into the library's website (2014). Responsive design may also affect wearable technologies, such as watches and virtual reality goggles. Viewports would scale down to 75 percent on watches, yet may expand to support virtual landscapes. This changes a number of factors, including user interactions, graphics, and content. However, it remains to be seen if libraries should develop for these emerging platforms, as they may not be the best way to deliver content to patrons.

The importance of crafting a satisfying user experience that places an emphasis on making human connections supersedes these newer trends. Mobile devices are becoming more ubiquitous, accounting for 30 percent of all Web traffic globally ("We are Social", 2016). Ten percent of Americans rely on smartphones for Internet access, especially minority and low-income users (Smith, 2015). This situation requires libraries to adopt new mobile technologies to serve these populations.

Analytics continues to play a significant role in the development of the library website, but it is not the only way to get insight into user behavior. That said, utilizing analytics information was useful for updating the mobile site's various iterations. For instance, we could see what mobile devices our users were utilizing to visit the site, and the site could then be catered specifically for this audience. Unfortunately, this has the reverse effect of not optimizing the site for users who connect with uncommon mobile devices. Perhaps by adhering to the appropriate web standards and guidelines, the library's responsive site can cater to these users as well. Yet, analytics is only one small piece in a larger puzzle. Usability tests and user feedback should be prioritized for future website improvements. Knowing how many users connect to the site with what device does not, on its own, indicate how users are interacting with the site.

# References

BBC News. (2014). Millions of websites hit by Drupal hack attack. Retrieved from http://www.bbc.com/news/technology-29846539.

Bevan, N., Barnum, C., Cockton, G., Nielsen, J., Spool, J., & Wixon, D. (2003). The magic number 5: Is it enough for web testing? In *CHI'03 Extended Abstracts on Human factors in Computing Systems* (pp. 698–699). ACM.

Black, E. L. (2011). Selecting a web content management system for an academic library Website. *Information Technology & Libraries, 30*(4), 185–189.

Bridges, L., Gascho Rempel, H., & Griggs, K. (2010). Making the case for a fully mobile library web site: From floor maps to the catalog. *Reference Services Review, 38*(2), 309–320.

Cummings, J., Merrill, A., & Borrelli, S. (2010). The use of handheld mobile devices: Their impact and implications for library services. *Library Hi Tech, 28*(1), 22–40.

Dresselhaus, A., & Shrode, F. (2012). Mobile technologies & academics: Do students use mobile technologies in their academic lives and are librarians ready to meet this challenge? *Information Technology and Libraries (Online), 31*(2), 82.

Faulkner, L. (2003). Beyond the five-user assumption: Benefits of increased sample sizes in usability testing. *Behavior Research Methods, Instruments, & Computers, 35*(3), 379–383.

Ganci, A., & McCullough, J. (2015). OCLC on the responsive web. *Library Technology Reports, 51*(7), 44–47.

Glassman, N. R., & Shen, P. (2014). One site fits all: Responsive web design. *Journal of Electronic Resources in Medical Libraries, 11*(2), 78–90.

Hubble, A., Murphy, D. A., & Perry, C. (2011). From static and stale to dynamic and collaborative: The Drupal difference. *Information Technology and Libraries, 30*(4), 190.

Kim, B. (2013). Responsive web design, discoverability, and mobile challenge. *Library Technology Reports, 49*(6), 29.

Lewis, C., & Rieman, J. (1993). Testing the design with users. In *Task-centered User Interface Design: A Practical Introduction* (http://hcibib.org/tcuid/tcuid.pdf).

Nielsen, J. (2000). Why You Only Need to Test with 5 Users. Retrieved from https://www.nngroup.com/articles/why-you-only-need-to-test-with-5-users/.

Reidsma, M. (2014). *Responsive Web Design for Libraries: LITA Guide*. Chicago: ALA Techsource.

Schofield, M. (2014). Bootstrap Responsibly [Web log post]. Retrieved from http://acrl.ala.org/techconnect/post/bootstrap-responsibly.

Sheehan, K. (2009). Creating open source conversation. *Computers in Libraries, 29*(2), 8–11.

Tidal, J. (2014). Testing on a tablet: Usability testing of a mobile library website. Presentation at the *Library Information Technology Association National Forum*, Albuquerque, NM, November 5–8.

Tidal, J. (2012). Using PHP to parse eBook resources from Drupal 6 to populate a mobile web page. *code4Lib Journal, 18*.

Tidal, J. (2011). Using Web Metric Software to Drive Mobile Website Development. *Computers in Libraries, 31*(3), 19–23.

We Are Social. (2016). Percentage of all global web pages served to mobile phones from 2009 to 2016. In Statista—The Statistics Portal. Retrieved from http://www.statista.com/statistics/241462/global-mobile-phone-website-traffic-share/.

Wilson, S., & McCarthy, G. (2010). The mobile university: From the library to the campus. *Reference Services Review, 38*(2), 214–32.

CHAPTER 3*

# A Mobile-First Library Site Redesign
## How Designing for Mobile Provides a Better User Experience for All

*Nathan E. Carlson, Alec Sonsteby, and Jennifer DeJonghe*

## Introduction

After nearly a decade of neglect and half-starts, Metropolitan State University's website needed an update. The university began a web redesign project to launch a completely renovated, mobile-first website that would reflect Metropolitan State's mission to deliver a vibrant, urban education for the Minneapolis-St. Paul, Minnesota, area and beyond. Although a contracted third party conducted the design work for the university homepage, much of the subsequent development happened in house, and the full site migration is still in progress at the time of writing. The library's own web team recognized an opportunity to utilize their knowledge of information-seeking behavior and universal design to improve the site's user experience for mobile users. Working in partnership with the university's Web Presence area and web developers in the IT Services department, the library web team was able to craft a stand-alone library homepage, fully functional on a smartphone, but sophisticated enough to satisfy researchers on any platform.

---

* This work is licensed under a Creative Commons Attribution 4.0 License, CC BY (https://creativecommons.org/licenses/by/4.0/).

# The Mobile Landscape
## *Where Users Are*

Increasingly, library websites must take multi-device users into consideration and move toward mobile-first site designs. Librarians may still be tempted to assume that library websites are too complex, and library research too involved, to be comfortably navigated with smartphones or tablets. The reality is that an increasing number of users are visiting library sites on mobile devices, and they do more than just check for hours or get directions. Indeed, a growing volume of evidence demonstrates that end users engage in complex tasks on mobile devices. A comScore report from March 2016 noted that 94% of 18-to-24-year-olds in the United States own smartphones and that an increasing percentage of mobile users are consuming books and video, handling transactions, and conducting research online. The same report showed that 62% of respondents access health information on a mobile device, and 32% of respondents access career information on mobile devices (comScore, 2016). The reason that many librarians assume that mobile users cannot do research on their phones might be rooted in the idea that only quick tasks can be done on a small device. This attitude is somewhat supported by studies such as the one conducted by Millward Brown Digital, which found that if a task is quick (five minutes or less), 81% of users in all age categories prefer to complete it on a phone. It is important to note, however, that even for tasks that last 10 to 20 minutes, 43% of respondents still preferred using a smartphone to using a laptop computer or tablet (2014). ComScore also found that recently, consumers have been choosing mobile devices with screen sizes of 4.5 inches or larger, and mobile phones with smaller screens are on the decline (2016), while evidence from other studies reinforces the idea that the larger the screen, the more time its owner spends using it (Wroblewski, 2016).

Librarians, who often interact with patrons who are visiting the library in person, tend to hold a skewed view of "typical" information-seeking behavior since many in-person visitors engage in longer, desktop-centric transactions. However, it should not be assumed that all research tasks take more than five minutes. Many information seekers make use of databases and research tools in short bursts: in 2015, 61% of e-book use "events" at Metropolitan State were sessions lasting five minutes or less, according to unpublished usage statistics. Additionally, some users may state a preference for using a desktop for research, but their actual behavior may deviate from that, particularly if they are in a location where mobile might simply be the most convenient available option. It is imperative that libraries recognize the role played by mobile devices in the lives of their users and the overall complexity of their online behavior.

## Accessibility and Equality of Access

Beyond recognizing the increased use of mobile devices, there are ethical reasons for concern regarding patrons who may not have the luxury of a choice of platform. In 2015, the Pew Research Center found that about 15% of the US population is considered to be "mobile-dependent," with either no other online access or limited online access. Additionally, those who are more likely to be mobile-dependent are disproportionately likely to be of lower socioeconomic status or from a nonwhite population (Pew, 2015). Neglecting mobile-dependent users means disregarding a significant percentage of the library user population. Many of these patrons are already in a more vulnerable position when it comes to equal access to library resources. Other mobile-dependent users are those with specific types of disability, for whom natural user interface (NUI) devices (touch screens, haptic feedback devices, and the like) are more adapted to their needs (Henning, 2016). The first item in the ALA Library Code of Ethics states that librarians should "… provide the highest level of service to all library users through appropriate and usefully organized resources; equitable service policies; equitable access; and accurate, unbiased, and courteous responses to all requests" (American Library Association, 2008). A mobile-first site design is necessary for providing equal access across the multitude of device types, patron types, and abilities.

## A Multi-Device, Mobile-First Approach

In a user experience (UX) framework, designers approach website organization by thinking about the interrelated journeys and tasks undertaken by users throughout their online environment (Greenberg, Carpendale, Marquardt, & Buxton, 2012). User journeys are discussed in detail later in this chapter; however, Paul Adams, a web developer who has worked for Google and Facebook, has also described how typical users are in fact multi-device users, navigating between a number of devices and screen sizes as they move throughout their day (2015). A user might very well begin working from a mobile device, switch to a desktop or laptop, and then transition back to mobile in the course of completing a single task. Thus, UX designers have moved from considering "mobile users" as a discrete group apart from "desktop users" and instead assume that both groups are moving between devices as a part of their user journey or story. Yet, as presented at a 2014 Event Apart conference, a mobile-first approach is still preferred when designing for multi-device users (Wroblewski, 2014). By designing for mobile first, and in particular for the smallest likely screen size, libraries can ensure that their sites present resources in the most usable and accessible manner. Since many sites designed for a desktop environment already suffer from clutter and an overly complex interface, designing for usability, accessibility, and simplicity on mobile delivers a better multi-device experience as well (Wroblewski, 2014).

Bad mobile design drives users away. In a 2012 study conducted by Google, 75% of US smartphone users surveyed preferred mobile-friendly sites, stayed on such sites longer, and returned to them more frequently. Conversely, 50% of users said that "even if they like a business, they will use them less often if the website isn't mobile-friendly" (Fisch, 2012, para. 5). The same Google study found that website features that users consider to be mobile "unfriendly" include slow loading time, small buttons and links, difficult-to-use input fields, and having to pinch and zoom (2012). Moreover, search engines like Google are now penalizing the ranking for sites that users do not consider "mobile-friendly" (Schwartz, 2016). Websites that are not mobile-friendly thus lose traffic in two ways: from users who simply avoid using the sites and from users who never find unfriendly sites because search engines suppress them.

### *Responsive Design*

The current recommended practice for designing for mobile is to use responsive web design (RWD), which is no less true for library websites (Tidal, 2015). In responsive web design, websites resize automatically based on the screen size of the device. This optimizes the experience for users whether they are viewing a site on a desktop, tablet, or smartphone. Given a multitude of mobile devices, from the smallest smartphones to large phones exceeding 5 inches, "phablets" of 5.5 inches and above, and full-sized tablets, all of which can shift from landscape to portrait mode, RWD has become not merely good practice but a necessity. This approach is preferable to creating a dedicated mobile site that is separate from a desktop site because a dedicated mobile site generally provides less content and a diminished experience for the user. A responsive site standardizes functionality across devices, an important consideration for multi-device users. Responsive web design is also preferable to the creation of mobile apps for institutions like libraries because of the effort and difficulty of maintenance and design consistency across multiple devices (Glassman & Shen, 2014).

# Understanding User Needs and Behaviors

## *User Demographics and Device Use at Metropolitan State University*

Metropolitan State Library's Google Analytics account currently reports an average of 23,356 unique monthly users. Just over 10% of those visitors accessed our site using a mobile device in 2015. While the total number of mobile users of the library website is still small as an overall percentage, that number has been rising steadily over the years. Additionally, the Metropolitan State University Library may report lower mobile usage than other libraries because the site has only recently become mobile-friendly and because of differences in the university's user population.

Metropolitan State has 11,505 currently enrolled students, 43% of whom are students of color. The average student age is 32 (Metropolitan State University, 2016). While the university does not keep statistics on the number of students who are mobile-dependent, given the demographic profile of the university it can safely be assumed that many students fit this category. In any given term, around one-third of courses offered at Metropolitan State are offered online. Students enroll from across the Minneapolis-St. Paul Metropolitan Area, outstate Minnesota, and the world. Thus, by the time of graduation, 97% of students will have taken at least one online or hybrid course (2015 data). This suggests that even students who may not consider themselves "online students" are very likely going to navigate an online learning experience at some point, and analysis of their device use is especially important.

## *User Surveys*

The intuitive way to understand user needs is to simply ask those users what their needs are. However, many usability experts caution against conducting surveys or focus groups, as users are notoriously bad at accurately reporting their behavior, and even worse at predicting what they will do in the future (Nielsen, 2001). In other words, what users say they do, and what they actually do, often diverge widely. Web design books such as *Rocket Surgery Made Easy* instead urge designers to rely primarily on usability testing, where one can observe user behavior in action and let actions and experiences, not stated preferences, guide web design (Krug, 2010). Despite their limited value, surveys provided the library team with some guidance during the early phases of the redesign, especially before the team had formed a working prototype with which to begin usability testing. At Metropolitan State University, the library web team conducted two surveys as part of the web redesign process. The first survey targeted library staff only, while the second (created and distributed with the help of the Web Presence area), focused on a broad spectrum of library patrons. In the survey of library staff, the questions centered on tasks that staff had observed end users struggling with when attempting to use the library website. The assumption was that many of the phone calls, instant messages, and in-person visits to service points in the library reflect usability faults with the website. While this assumption is an obvious oversimplification, it provided some useful starting points from which the library web team could begin the process of site design. In the survey that went out to end users, the questions asked about the reasons respondents visited the library website, the types of devices used, and their overall satisfaction with the site for performing various tasks. (See appendix 3A. User Survey.)

Both surveys conducted as part of this process were created using the free version of SurveyMonkey. The staff survey link was sent via e-mail, and the survey of end users was shared via e-mail and linked on the website and through social media. The 257 respondents for the broader survey were likely not representative of the user base as a whole as there would be a self-selection bias. Nevertheless, the survey instrument was free and yielded useful information.

## *User Personas and Job Stories*

While surveys hint at which features users may want on the library's homepage, and analytics confirm actual traffic patterns, neither of these tools reveal information about types of functionality that users do not yet know that they want or that does not yet exist. Furthermore, the usual model of presenting every library tool at the same time often obscures the most common functions and overwhelms users. The creation of personas and job stories allowed the web team to personalize the experience of the library's users, explaining how they "will use, experience value in using, and continue to use" the site (Lichaw, 2016, p. 70).

The library web team had two user personas developed by graduate students as part of a course project in the technical communications program at the university. The personas were designed to represent two different types of students that staff commonly see at the library, one being a tech-savvy mobile user who is a first-generation immigrant, and the other an older returning adult student who also uses mobile but in a different way, focusing on transactions rather than social engagement. The web team uses personas both to help visualize the library's end users and to establish context for scenario mapping or creating job stories. While the personas represent fictional users, they assist the team in staying grounded in the practical realities end users face, in identifying "pain points" where users get stuck trying to meet their needs, and in maintaining a higher level of empathy.

With personas established, the development team created job stories that illustrate the tasks users need to accomplish. The team identified a job, broke down that job into its constituent tasks, and described how each persona might complete these tasks on the current site, which helped to illuminate the causality and motivations for the tasks. Finally, the team brainstormed changes to the current system that would improve the job and satisfy the user context. For instance, one job that students face is finding time in a busy schedule for studying. The subtasks might include finding out when the library is open, reserving a study room in the library, or finding study materials quickly. One of the personas, "Joe," has a smartphone and a laptop computer and works full-time during the day. Joe encounters the complicated tasks of finding the library's hours, reserving a study room, or using his smartphone to navigate the website. These tasks were irritatingly difficult on the previous, mobile-unfriendly website. The development team might implement such solutions as listing the current hours on the homepage or creating mobile-friendly study room booking and course reserves modules. By putting themselves in Joe's shoes, the team could identify the quickest and most direct "path" for a user navigating the site. The user journey maps were created on a whiteboard with colored markers. Boxes represented tasks, and arrows illustrated the user's course of action. The team photographed the maps and referred to them throughout later design phases.

# Pre-Design Phase
## *Collaboration*

A key part of the web design work done at the Metropolitan State University Library is based on a collaborative model and team-based approach, both within the library and between the library and external departments. Since its inception, the library web team has been comprised of a mix of librarians and library staff from both public services and technical services. The web team has a team charter and a "usability guiding document," written collaboratively and reviewed annually as a way to foster a shared philosophy of user-centered web design. The charter specifies that no particular skill set is required to join the team and that members need only to be enthusiastic and dedicated to improving the library web presence. Group work and meetings are often done in a flexible work space (a simple room with a table and a stand-up computer or a smart conference room), and work is documented in a wiki on PBworks that is shared with the entire library department.

Metropolitan State University staff operate within a complex union environment, with a total of five unions and a number of university departments scattered across multiple physical campuses. The ownership and governance of the university website is a frequent source of formal and informal debate, and ongoing friction over whether the website should be "housed" in the Marketing department, the IT Services department, or elsewhere has slowed development of the university's site redesign process. Librarians sit on the university-wide Web Advisory Committee and strive to enhance the university's approach to web presence and to push for broader discussion about digital issues.

The web redesign team for the university consists of members from the IT Services department (ITS) as well as the Web Presence Director, who currently works outside of ITS in a department that also includes Marketing and Publications. The Web Presence team and ITS connect with frequent scrum meetings, short, daily meetings intended to drive development teams forward toward a common goal. Library staff have separate meetings at less frequent intervals and meet together with the university web redesign team about twice a month.

As is the case at many institutions of higher education, the library often struggles to maintain control over the library website content and to maintain that control during and through times of institutional change and staff turnover within other departments. As a result, the library web team has pursued, through the use of frequent meetings and communications, close working relationships with individuals in external departments who can work with, and advocate for, the librarians. Because the library recognizes and appreciates the talented developers, strategists, and UX professionals in the other departments, those staff are, in turn, more willing to work with us. And the library often has resources to contribute, whether it be people to share in the work, hardware for usability testing, or budgetary room to share in the cost of software licenses. The collaborative library's toolkit therefore should include humility, gratitude, and altruism.

### *Collaboration Tools*

While frequent in-person meetings between the library and external departments are logistically difficult, in between the regular meetings librarians frequently collaborate and communicate with each other, and external departments, using a variety of online tools. While working on the site content, librarians frequently chatted with ITS developers and the Web Presence Director using Skype for Business and Facebook Messenger. Additionally, a Basecamp 3 collaborative space was established for sharing materials and files with the full team. In some cases, the collaboration tools used were simply what was most convenient at the time, with no sense that use would be ongoing or permanent. Moving from one communication and collaboration tool to another has been of little concern. The particular tool used is the least important aspect of team collaboration—an underlying philosophy of rapid and transparent communication powers the success of cross-functional teams.

## Site Design and Testing

The university's broader web redesign plan originally included collaboration with, and ultimately a homepage design from, a local web development firm selected through an RFP process. Because this plan and timeline changed (the project ran over budget and the development firm did not design the entire university site), the library web team developed a timeline of its own. With a highly complex site and the liberty of time due to the content migration delay, the library web team embraced an iterative design philosophy. This included a cyclical process, informed by competitor design analysis, usability testing, wireframing, card sorting, and mock-up. Many design projects employ these steps in succession, moving from analysis to card sorts to wireframing to usability testing to final design (Rosenfeld, Morville, & Arango, 2015; Morville, 2014). While this methodology can shorten a site's time-to-launch, hastening the process can cause a design team to overlook moments of serendipitous discovery and chance findings. Providing structure while allowing space for spontaneity can result in a richer design.

## *Usability Testing*

In response to feedback that the last website redesign at Metropolitan State University had launched with insufficient input from actual users, the library web team felt that a successful library webpage should incorporate observations of user behavior into the design cycle. The library has a practice of conducting usability tests at a minimum of once per semester. Mobile behavior has been harder to capture because it involves new strategies and equipment that disrupt old models of usability testing. In some forms of mobile usability testing, the team provides a mobile device that is preloaded with capturing software to be used by test participants. However, purchasing a piece of hardware in a state university requires the intervention and approval of several departments, including the library, Purchasing, ITS, and Accounts Payable departments. The library web team could not undertake this process every 14 months or so to keep up with the latest phone and software releases, in addition to managing the complexity of sharing a single device tied to an individual with a cross-functional web team. Furthermore, having test subjects use phones and operating systems they are not familiar with can be disorienting for them (Cerejo, 2016). Despite these complications, with the university launching a mobile-first site the web team could no longer ignore the importance of capturing user behavior on mobile devices. The team adopted a practice of mobile testing where users are observed using their own devices, which are positioned below a document camera. The camera records the test and projects it on a screen behind the participant for the observers to see.

There are a number of methods for usability testing, but the library web team employs a model adapted from the book *Rocket Surgery Made Easy* by Steve Krug (2010), which espouses a strategy of conducting frequent, simple, and inexpensive usability tests. Following this model, participants are recruited from the main library study area on the day of the testing, invited to help with improving the library's website, and offered an incentive of a gift card for five dollars from the university's bookstore. The participant is then shown into a separate room where the library web team has set up a laptop and a second screen that mirrors what is on the laptop. One member of the team reads through a script and guides the participant through a predetermined set of tasks (see appendix 3B. Test Script) while other team members observe and record notes about the participants' movements as seen on the second screen. Participants are encouraged to talk through their thoughts and actions in order to provide additional context for their on-screen behavior. After the test concludes, the team debriefs by comparing notes and impressions. Later, the team organizes the different sets of notes into a shared document and lists any action items resulting from the test results. For the homepage design project, the library web team sought and obtained formal approval of its usability testing from the university's Human Subjects Review Board (HSRB). While this approval was unlikely to be necessary, since the results from usability testing are for internal use, the web team pursued it as an extra precaution to cover the potential use of data in future publication.

## *Wireframing*

Because coding an entire prototype in HTML and CSS within the university's content management system would have been time-consuming and effort-intensive, the web team first sketched on paper and whiteboard, then used prototyping software, to create simple wireframe models. The web team, following Rosenfeld, Morville, and Arango, needed a way to represent conceptual architecture in an interactive way (2015). Wireframes, mock-up sites with minimal styling and functionality, served as a way to test the site's layout and architecture without fully coding details that may still have been in flux.

The web team embraced a form of rapid prototyping based on wireframing software Axure. Axure's drag-and-drop graphical layout editor allowed the team to quickly develop wireframes, make iterative changes, incorporate team feedback, and share a testable site with the university web developers and test participants, all within a matter of hours. This process was so convenient, in fact, that the team was able to record feedback from one usability test participant, modify the prototype, and test these changes with the next participants within the same testing session. The Axure interface also enabled the team to create a responsive layout that displays correctly on various device widths, a crucial detail for testing on participant-provided devices.

The team conducted two rounds of wireframing and tested them each with users: an A/B test (figure 3.1), followed later by a test with a functional mock-up (figure 3.2). The first round became an A/B test because one member of the development team was certain that users would prefer links listed in a drop-down menu, supported by the theory that users dislike scrolling to find content. Other members of the team looked at competitor designs and found few drop-down menus but lots of content boxes, lists of links, or other content set off from the rest of the page by a contrasting box, suggesting a general movement away from drop-down menus. For the A/B test, then, the web team constructed a wireframe with each format and asked participants to accomplish tasks using first version A, then version B, or vice versa. Test participants overwhelmingly preferred the content boxes, and, more importantly, accomplished the tasks more quickly with that model.

After the A/B test, a new mockup was created that used the content boxes and added full homepage functionality, including a more refined layout, input boxes, and clickable links. This second round of wireframing not only confirmed the results of the card sort (discussed below), but also revealed the comfort level that users have with the new site. For example, when test participants were asked to locate a link to a study-room booking tool, two users began to reserve study rooms for themselves during the test. Another user had trouble locating the library's hours, listed at the top of the page, and instead scrolled all the way to the bottom where she expected to find them, suggesting that scrolling is not a barrier for users.

**FIGURE 3.1**
Wireframe A/B test made in Axure.

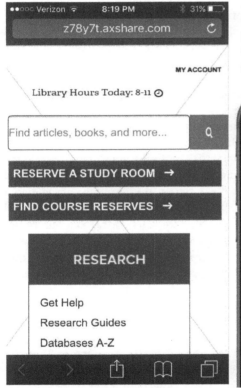

**FIGURE 3.2**
Functional homepage design wireframe made in Axure, displayed on iPhone.

The wireframe results influenced not just the final design, but intermediate design steps as well. Clarity on the question of scrolling versus drop-down menus persuaded the team to include more links and information broken out on a longer homepage (which takes more time to scroll through), rather than hidden in menus. Many test participants indicated that instead of using links or a site-specific search function, they would use Google to find the Metropolitan State library information they wanted and bypass the homepage navigation entirely, an attitude that spawned a fruitful subproject to improve the library's search engine optimization and Google search result rank. Finally, as Rosenfeld et al., point out, a wireframe "helps clarify the grouping of content components, their order, and group priority" (2015, p. 407). This, combined with feedback on the A/B wireframes, led the web team to rethink the labels and link names used in the card-sorting activities.

## *Card Sorting*

Concurrent with the wireframing and usability testing, the library conducted card-sorting sessions both physically, with notecards and pens, and digitally, using a program called Optimal Sort, which is part of Optimal Workshop (figure 3.3). Participants took prelabeled cards representing existing content pages and arranged them by type, then named the new categories using whatever language seemed appropriate. For the physical card sort, two populations participated: library staff, and a classroom of undergraduate information studies students. Staff are expert users, and the information studies students were moderate to expert users since they had spent most of a semester using the library website. The library web team did not intentionally seek out expert users, but used a population that was readily available. Thus, the team had to consider that the users who participated were atypical users and factor that into their discussion of the results. In the digital card sort, the library web team created the digital "cards" but again allowed the participants to name the categories themselves. The advantage of the digital card sort was that the library was able to solicit participants via e-mail and social media who were remote to the library, including students at local community colleges who were not enrolled at Metropolitan State University and thus had no familiarity with the old website. The advantage of the physical card sort came from observing the participants, watching their facial expressions, and noting the time it took to complete the task. In some cases, participants articulated confusion over language used for particular cards, which we noted for follow up testing in other forms of study, for example, usability testing.

From the card-sorting activities, the library web team gleaned information that helped with the overall mapping of the site and the site architecture. The team got a sense for how end users logically clustered activities and information together, often in ways that deviated from the expectations of librarians. For example, most end users sorted "Interlibrary Loan / Article Delivery" into categories called "Research" whereas library staff often placed it into a generic "Services" category. Card sorting also revealed what page content students deemed unimportant. Pages of library policies, information about specific programs, and the staff directory were left unsorted or put into a category called Other. The response to the card-sorting activity was positive overall, and participants reported that they enjoyed completing the activity.

**FIGURE 3.3**
Screenshot of digital card sort in Optimal Workshop.

## Challenges and Future Directions

Metropolitan State University faces a considerable challenge due to the lack of time and resources dedicated to web development and usability testing. No one library staff member has been assigned the development, maintenance, and evaluation of the website as a primary duty. Web design consists of work that the team of staff and librarians do when they can fit it into their schedules and often at the expense of other, primary duties.

The way patrons approach websites has changed rapidly in the past ten years as mobile device use has exploded and device types continue to proliferate. Given that environment, the lack of dedicated resources and support for professional development opportunities at Metropolitan State University has made it difficult to keep pace. This makes continued collaboration with staff outside of the library (such as with ITS) imperative. Many levels of university administration continue to exhibit twin blind spots in understanding the role of technology in students' lives and in understanding their obligation to provide a true digital campus. This perception gap results in organizational decisions that hinder the work of web developers, librarians, and staff.

It is the desire of the university's Web Presence Director, as well as the library web team, to avoid any further "redesigns" and to move forward with a philosophy of continuous improvement. Once the library site redesign goes live, the next steps will center on continued usability testing to fine-tune a user-centered experience for patrons. There is much that still can be improved, but with a mobile-first foundation in place, the site can be adjusted and improved in smaller, incremental steps. Such steps will include targeted usability testing that focuses more specifically on users with disabilities to ensure that the design is as universally usable as intended. Additionally, the librarians intend to pursue a slightly more formal process of usability testing. While the light, "pop-up" style of usability testing has a number of advantages already described, the results are difficult to share in a way that is meaningful for people outside of the web team. Polished-looking, shareable results can help garner buy-in from decision-makers at the university, which is especially important when results could point to a need for investment in additional personnel and resources. Sharing results also illustrates to administrators the amount of behind-the-scenes work that goes into designing a good website. Finally, the impact of watching and listening to an end user as he or she struggles to accomplish tasks cannot be overstated and goes a long way toward convincing stakeholders of the importance of a highly usable website. Thus, a process using screen-recording and annotation software will be pursued. These recorded sessions will be presented in staff meetings or shared over e-mail both within the library and to the broader university community.

As the site design becomes more established, the library can broaden and deepen testing by creating additional user personas and more nuanced user journeys. For example, the team can probe the question of whether certain users engage in forms of quicker research on a phone and then transition to a desktop or laptop for longer tasks and discuss how that might inform our design. End users demonstrate complex behavior, using the library in a multitude of ways on a multitude of devices in the course of their day and in their academic career. The library can facilitate their successful navigation of this environment using personas, user journey mapping, and usability testing and meet that complexity with simplicity in site design.

# Acknowledgments

The authors would like to thank Steven Adrian, university Web Presence Director, and his team; Richard Harrison, Senior Interactive Designer; and the entire library web and usability team.

APPENDIX 3A
# Metropolitan State Library Homepage Redesign, User Survey

1.  Are you a(n):
    a.  Undergraduate Student
    b.  Graduate Student
    c.  Faculty/Staff
    d.  Alumni
    e.  Community Member
    f.  Other (please specify)
    **[checkbox, with Other being open, multi select]**

2.  During the school year, I use the Library website:
    a.  Daily
    b.  Weekly
    c.  Monthly
    d.  Once a semester
    e.  Never
    f.  Other (please specify)
    **[Radio button, with Other being open, limit to one selection]**

3.  From what location do you usually access the Library website?
    a.  Library
    b.  From another location on campus (Saint Paul, Minneapolis, Midway or Brooklyn Park)
    c.  Work
    d.  Home
    e.  I don't usually visit the Library website
    f.  Other (please specify)
    **[Radio button, with Other being open, limit to one selection]**

4.  What is the **most common** reason you use the Library website?
    a.  Finding resources related to research or course requirements
    b.  Checking my library account, renewing items
    c.  Getting directions or hours for visiting the library in person
    d.  Seeking assistance from a librarian, via phone, chat, or reference desk
    e.  Get information on citing sources (APA or MLA)
    f.  Getting e-reserves associated with a class
    g.  Other (please specify)
    **[Radio button, with Other being open, limit to one selection]**

5. When you visit the Library website, how often do you do the following:
   a. Find items physically in the library building, such as books, DVDs, and other items
   b. Renew checked-out items
   c. Access e-reserves
   d. Request materials from other libraries using Interlibrary Loan
   e. Use the library One Search (search box), to research topics
   f. Use the Library's databases (such as Academic Search Premier, Lexis-Nexis, etc.) to locate an article from a journal, magazine or newspaper on a specific topic
   g. Access e-books
   h. Check library hours
   i. Find contact information for Library staff
   j. Ask questions (Ask a Librarian chat or email)
   k. Get help with citing sources and formatting a bibliography (such as APA or MLA)
   l. Use tutorials and guides related to academic papers or technology
   m. Find a specific electronic magazine or journal by name
   n. Use electronic encyclopedias/ e-reference
   **[Scale Headings: often, sometimes, rarely, never, did not know I could do this]**

6. Which statement best reflects your experience when navigating the Library website?
   a. I am always able to find the services, databases or resources I look for.
   b. I am usually able to find the services, databases or resources I look for.
   c. I have difficulty finding the services, databases or resources I look for.
   d. I am unable to find the services, databases or resources I look for.
   e. Other (please specify)
   **[Radio button, with Other being open, limit to one selection]**

7. Would you be willing to participate in a face-to-face study to improve the usability of the Library website?
   a. Yes
   b. No
   c. Other (please specify)
   If "Yes," please provide your full name and email address.
   **[Radio button, with Other being open, limit to one selection]**

8. Are there any other services or features you would like to see added to the Library website?
   **[Open-Ended Question]**

APPENDIX 3B

# Metropolitan State Library Homepage Redesign, Test Script: Wireframe 2

## *Introduction*

Hi, [participant name], my name is [your name], and I'm going to be walking you through this session.

You've already heard a little bit about this, but let me explain more about why we've asked you to come here today: We're testing the Metropolitan State University Library's website to see what it's like for students and library patrons to use it.

I want to make it clear right away that we're testing the site, not you. You cannot do anything wrong here. You may quit your participation at any time and it won't affect your relationship with me, or anyone else at Metropolitan State University.

We want to hear exactly what you think, so please don't worry that you're going to hurt our feelings. We want to improve the Library's website, so we need to know honestly what you think.

As we go along, I'm going to ask you to think out loud, to tell me what's going through your mind. This will help us.

If you have questions, just ask. I may not be able to answer them right away, since we're interested in how people do when they don't have someone sitting next to them, but I will try to answer any questions you still have when we're done.

Do you have any questions before we begin?

## *Background Information Questions*

Before we look at the site, I'd like to ask you just a few quick questions. First, how long have you been a student at Metro State?

Good. About how many hours a day would you say you spend using the Internet, including email?

Have you used the Library's website before? [If yes] how often? On what kind of device (desktop, phone, tablet)? [If no] Why not?

OK, great. Now we can start looking at the library website.

### *Usability Test*

1. [Start at library homepage.] You're on the way to meet your study group at the library, but you heard the library was closed. How would you check to see if the library is open? How would you find the hours for tomorrow? [desired response: finds hours]
2. You were supposed to book a room for your study group, but you forgot until now. Where would you go to book a room? [desired response: finds link to study room booking]
3. You are supposed to research for a paper, but you have never done this before. Where would you go to get started with your research? [desired info: student finds Library Guides or OneSearch]
4. This page is totally confusing! How would you get help? [desired: student clicks chat widget or get help]
5. You have had a book checked out for weeks and it's overdue! Where would you go to renew it? [desired info: student clicks My Account or Checkouts and Renewals]
6. If you wanted to get a copy of a book or journal article that Metro library doesn't have, what would you do? [desired info: finds interlibrary loan link]
7. How long can you check out a DVD from the Metro Library? [desired info: finds policies or checkouts and renewals]
8. Can you think of other reasons you might check the library website? How would you find that here?
9. Any other comments on this page?

Thank you, that was extremely helpful. Do you have any questions for us, now that we are done?

# References

Adams, P. (2015, May 13). Why 'mobile first' may already be outdated. Retrieved May 11, 2016, from https://blog.intercom.io/why-mobile-first-may-already-be-outdated/.

American Library Association. (2008, January 22). *Code of ethics of the American Library Association*. Retrieved May 1, 2016, from http://www.ala.org/advocacy/proethics/codeofethics/codeethics.

Cerejo, L. (2016, February 24). Noah's transition to mobile usability testing. Retrieved May 11, 2016, from https://www.smashingmagazine.com/2016/02/mobile-usability-testing/.

comScore. (2016, March 30). 2016 U.S. cross-platform future in focus. Retrieved May 1, 2016, from http://www.comscore.com/Insights/Presentations-and-Whitepapers/2016/2016-US-Cross-Platform-Future-in-Focus.

Fisch, M. (2012, September 25). Mobile-friendly sites turn visitors into customers. Retrieved from http://googlemobileads.blogspot.com/2012/09/mobile-friendly-sites-turn-visitors.html.

Glassman, N. R., & Shen, P. (2014). One site fits all: Responsive web design. *Journal of Electronic Resources in Medical Libraries, 11*(2), 78–90. doi:10.1080/15424065.2014.908347.

Google, Sterling Research and SmithGeiger. (2012). What users want most from mobile sites today.

Greenberg, S., Carpendale, S., Marquardt, N., & Buxton, B. (2012). *Sketching user experiences: The workbook*. San Francisco, CA: Morgan Kaufmann.

Henning, N. (2016). Mobile learning trends: Accessibility, ecosystems, content creation. *Library Technology Reports, 52*(3), 1–38.

Krug, S. (2010). *Rocket surgery made easy: The do-it-yourself guide to finding and fixing usability problems*. Berkeley, CA: New Riders.

Lichaw, D. (2016). *The user's journey: Storymapping products that people love*. Brooklyn, NY: Rosenfeld Media.

Metropolitan State University. (2016). Key facts at a glance. Retrieved May 10, 2016, from http://www.metrostate.edu/why-metro/about-the-university/key-facts.

Millward Brown Digital. (2014). Getting audiences right: Marketing to the right generation on the right screen. Retrieved May 10, 2016, from https://www.millwardbrowndigital.com/Research/getting-audiences-right/.

Morville, P. (2014). *Intertwingled: Information changes everything*. Ann Arbor, MI: Semantic Studios.

Nielsen, J. (2001, August 5). First rule of usability? Don't listen to users. Retrieved May 13, 2016, from https://www.nngroup.com/articles/first-rule-of-usability-dont-listen-to-users/.

Pew Research Center. (2015, April 1). *U.S. smartphone use in 2015*. Retrieved from http://www.pewinternet.org/2015/04/01/us-smartphone-use-in-2015/.

Rosenfeld, L., Morville, P., & Arango, J. (2015). *Information architecture: For the web and beyond*. Sebastopol, CA: O'Reilly.

Schwartz, B. (2016, May 12). Google's mobile-friendly algorithm boost has rolled out. Retrieved May 12, 2016, from http://searchengineland.com/googles-mobile-friendly-algorithm-boost-rolled-249357.

Tidal, J. (2015). One site to rule them all: Usability testing of a responsively designed library website. *Proceedings of the 2015 Association for College and Research Libraries Conference*, Portland, OR. Retrieved May 12, 2016, from http://www.ala.org/acrl/acrl/conferences/acrl2015/papers.

Wroblewski, L. (2014, September 23). An event apart: Mobile first responsive design. Retrieved May 8, 2016, from http://www.lukew.com/ff/entry.asp?1923.

Wroblewski, L. (2016, January 25). As mobile screen size increases… So does activity. Retrieved May 7, 2016, from http://www.lukew.com/ff/entry.asp?1956.

CHAPTER 4

# Selfie as Guide
## Using Mobile Devices to Promote Active Learning and Student Engagement

*Sarah LeMire, Stacy Gilbert, Stephanie Graves, and Tiana Faultry-Okonkwo*

## Introduction

Engagement is a big buzzword on college and university campuses these days, and libraries are no exception. While librarians frequently turn to pedagogical techniques such as active learning in order to increase student engagement during information literacy instruction sessions, far less has been written about transforming an old standby: the library tour. Librarians at Texas A&M University took advantage of an unexpected opportunity to acquire mobile devices in order to develop a pilot "selfie-guided" tour that moved librarians out of the "sage on the stage" tour guide role. Instead, this new tour required students to form teams and use library-provided mobile devices to tour the library on their own, answering questions about different library spaces and services, and to take selfies in different library spaces, which were then shared with the class. The selfie-guided tours challenged students to navigate library spaces on their own, to engage actively with library staff in order to answer their assigned questions successfully, and to work together and use their creativity in taking selfies.

# Library Tours as Academic Library Outreach

Library tours are a popular method of introducing students to academic libraries. By giving users a broad overview of the library's physical layout, website, and collections, library tours can help orient students to the library's space and resources (Ingalls, 2015). Numerous articles have cited the benefits of tours in academic libraries, with the most frequently discussed benefit being the familiarization of students with the physical layout of the library's facilities, collections, and resources (Foley & Bertel, 2015; Kearns, 2010; Sandy, Krishnamurthy & Rau, 2009; Sciammarella & Fernandes, 2007). Tours make students aware of library services in context of their physical location, and students can practice seeking assistance from library staff (Foley & Bertel, 2015; Kearns, 2010; Sciammarella & Fernandes, 2007). It has also been noted that tours that take advantage of technology can demonstrate for students how to find online library materials and databases (Kearns, 2010). Understanding the physical layout of the library, and its collections and services, demonstrates to students how the library can assist them in their collegiate studies. This can make students aware of the resources available to them at the early stages of the research process and also help them develop a foundation on which to build additional research skills (Kearns, 2010; Sciammarella & Fernandes, 2007).

A tour can also have a positive impact on students' perception of the library. The tour might be a student's first encounter with an academic library and can reduce anxiety related to using the library (Cairns & Dean, 2009). Marcus and Beck (2003) found that after students at Queensborough Community College of the City University of New York took a tour, they viewed the library staff as helpful and enthusiastic and the library atmosphere as pleasant and conducive to studying. Showing students a library's welcoming, friendly environment helps reduce anxiety by demonstrating practical applications of how the library can assist students with their studies.

# Active Learning and Mobile Technology

Tours are typically passive learning experiences; the tour guide speaks while participants listen. However, learning theory shows that passive learning isn't always the best strategy for retaining information. Instead, active learning is commonly described in the literature as a successful pedagogical alternative. Reports dating back to the 1980s and 1990s note that, "more learning occurs when students are actively engaged in the learning process" (National Institute, 1984, p. 19) and suggest strategies for incorporating active learning into the college classroom (Bonwell & Eison, 1991). More recent research continues to confirm that active learning techniques result in improved instructional outcomes in a variety of disciplines (Prince, 2004; Michael, 2006; Freeman et al., 2014). Academic librarians have also embraced active learning, and research has confirmed the efficacy of active learning as a technique for improving library instruction outcomes. Drueke (1992) found that "students in the active learning classes incorporated more relevant library research into their final projects" (p. 82). Detlor, Booker, Serenko, and Julien (2012) confirm that "active [information literacy instruction] has a direct effect on yielding positive student learning outcomes, while passive [information literacy instruction] does not," and they suggest that librarians may want to "limit or even eliminate the delivery of passive [information literacy instruction] altogether" (p. 156).

Many academic librarians have enthusiastically adopted active learning as a pedagogical tool and developed activities to engage students in the classroom, as evidenced by the proliferation of books that collect library instructional activities (Sittler & Cook, 2009; Cook & Sittler, 2008; Fawley & Krysak, 2016). Recently, librarians have also been harnessing the capabilities of mobile technology in order to develop new active learning strategies to engage library users. Several librarians have reported success using clickers or mobile phones to elicit student responses and heighten student engagement (Burkhardt & Cohen, 2012; Hoppenfeld, 2012). Other librarians have used iPads in reference transactions in order to provide students with the opportunity to work directly with library resources and therefore increase student engagement (Maloney & Wells, 2012). Additionally, librarians have deployed iPads and tablets in library instruction in order to encourage students to engage with library resources (Havelka, 2013; Gilbeault, 2015), practice concept mapping (Calkins & Bowles-Terry, 2013), or participate in scavenger hunts (Miller & Putnam, 2015).

Mobile devices also have the potential to help students connect with information as creators, not just consumers. Lippincott (2010) astutely notes:

> As librarians work with students as part of information literacy classes, at service desks, and in cyberspace, it is important to realize that for students, the mobile device will increasingly become an instrument for creation of digital content, and not just a device for access to content. Students can use smartphones to create short videos, to type a blog entry for a class assignment, to "tweet" in response to a question posed by a professor or to create a group poem, or to take photos or record audio to embed in a PowerPoint presentation or text document (p. 210).

Mobile devices can be used as a tool to engage students in active learning even when participating in traditionally passive learning activities such as library tours.

# Mobile Technology and Library Tours

Self-guided tours of academic libraries have been around for more than forty years. As summarized by Oling and Mach (2002), self-guided tours using printed booklets gained popularity among academic research libraries in the late 1960s and early 1970s. Technologies evolved, and self-guided tours began to incorporate websites with virtual tours in the 2000s. Early virtual tours allowed the user to tour the library from a stationary computer. Today's self-guided tours are incorporating mobile technologies like MP3 players and tablets, which allow guests to learn about the library while moving through the physical spaces. When one is researching self-guided tours that incorporate mobile technologies, two types of tours emerge: the first type is unmediated and available to be taken by the student anytime, and the second type is game-based tours.

Unmediated and asynchronous tours typically involve audio or video tours that students access via an MP3 player or tablet. The majority of these tours can be taken whenever the student is free and without a librarian facilitating the tour, freeing up time for librarians to complete other activities. Virginia Tech and Brigham Young University (BYU) have utilized Quick Response (QR) codes to create library tours. QR codes are a type of mobile-ready barcode that launches a website, audio file, or application on the mobile device. Students use their mobile device to take pictures of the QR codes to access audio or video files that explain the different service points or resources in the library. Virginia Tech and BYU also provide post-tour quizzes for students to complete. As BYU's mobile tour evolved, the audio files were replaced with video files accessible with iPods (Virginia Tech, n.d.; Whitchurch, 2015). The iPods were later replaced with affixed iPads "programmed to show the video for a specific tour stop/library location" (Whitchurch, 2015, Mobile to Affixed iPad section, para. 1).

Besides QR codes, some universities use audio or video files downloaded onto mobile devices. The University of Sheffield, the University of Alabama (UA), the University of Tennessee at Chattanooga (UTC), the University of South Carolina Upstate (USC Upstate), and the University of California, Merced (UCM) created video or audio recordings for patrons to listen to as they toured the library. For some of these tours, markers were placed around the library to inform students where they should stop and listen to a particular recording. The University of Sheffield and USC Upstate each wrote a script in which a student interacts with different librarians or library staff members, giving the audio a conversational feel. Most of these schools purchased iPods for students to check out and use for the tour, although UA purchased Sony Video MP3 media players. UTC, USC Upstate, and UCM targeted their tours to first-year student programs, and they also created a posttest quiz to assess students' familiarity with the library (Cairns & Dean, 2009; Kearns, 2010; Mawson, 2007; Mikkelsen & Davidson, 2011; Sandy et al., 2009).

Game-based tours typically take the form of a scavenger hunt or treasure hunt. These tours have been implemented at State University of New York at Buffalo (SUNY Buffalo) and North Carolina State University (NCSU). Students take photographs of different areas in the library with an iPad or iPod Touch. At SUNY Buffalo, early versions of the tour required students to upload photos to a common Flickr account, while later iterations of the tour used GooseChase, a free photo-based scavenger hunt app. At the end of the tour at SUNY Buffalo, the students described the photos to the class. NSCU's students submitted their text and photo responses using a shared Evernote account, and the librarians graded the scores using a Google Docs spreadsheet (Burke, 2012; Foley & Bertel, 2015; North Carolina State University Libraries, 2014).

Additionally, some universities used a mobile app called SCVNGR to aid their orientation tours. SCVNGR was a social location-based gaming app for iPhones and Androids. Users had to be in a location to search for activities and challenges and then perform those challenges in the specific location to earn points or recognition. SCVNGR is dissimilar to a scavenger hunt because "rather than getting clues and hints like in a traditional scavenger hunt, this game is more focused on activities within a location instead of finding the location" (Pagowsky, 2013, Why SCVNGR? section, para.1). The University of Arizona (UAZ), University of California, San Diego (UCSD), Oregon State University (OSU), Boise State University (BSU), and UCM all experimented with SCVNGR. Some of these schools, including UAZ, UCSD, and UCM, used SCVNGR for large-scale, campus-wide orientations. This app was good for large programs because it did not require librarians and library staff to commit a great deal of time to help with the tours. OSU used SCVNGR for its international student orientations. As part of an assignment at BSU, students were asked to create their own SCVNGR orientation. UAZ also used SCVNGR for individual classes (McMunn-Tetangco, 2013; Pagowsky, 2013). Today, SCVNGR is called LevelUp, a mobile-payment application, and no longer appears to support the location-based gaming features that made it popular with libraries for orientation tours (LevelUp, n.d.).

The project described in this chapter adds to the current landscape in several ways. First, most mobile technology tours are designed for library orientations outside of the normal classroom experience. There are a few examples in the literature of incorporating a virtual library tour in an information literacy one-shot class, but many struggle with the limitations of the traditional fifty-minute session. This project aims to show that thoughtfully created library tours can meet learning outcomes for one-shot classes. Additionally, the project focuses on the use of mobile technology to increase student engagement in a particular population: academically at-risk students. Thinking carefully about the specific student audience, their learning needs, and their challenges helped the authors design an interactive library tour customized for a specific population.

# Developing the Selfie-Guided Tour

The Texas A&M University Libraries provide numerous tours of its five libraries to various college classes, high school groups, and other community groups. The main library, the Sterling C. Evans Library and Annex, receives the bulk of the tour requests, which are coordinated by the University Libraries' Learning and Outreach (L&O) department. As library staff deliver these tours, there is an overall goal of iterative improvement for all library tours. Feedback is collected after every tour for the purposes of programmatic improvement, as insights gained from one tour group are used in planning future tours for other groups.

In the summer of 2014, L&O modified its traditional, in-person, staff-led tours in order to provide the students with a more impactful experience. L&O collaborated with the University Writing Center, Maps and GIS Collections, and Media and Reserves staff to implement stations that were tour stopping points, loosely modeled after the concept of learning centers. During the course of each tour, the library staff member serving as tour guide would take the group around the library and make a stop at each station. The staff members at that station would give students a personalized welcome, a brief presentation, and sometimes a "show and tell" of their collections and resources. The tour guide provided the information for all of the other library resources and service points covered by the tour. This modification encouraged more interaction and familiarity between students and staff at those individual service points within the library. However, students were still passive recipients of information for much of the tour.

In 2013 and 2014, L&O also began to experiment with strategies for integrating mobile technology into its library tours. L&O staff used six first-generation iPod Touches for scavenger hunt tours for the library's Academic Integrity curriculum and for some high school groups (Texas A&M University Libraries, 2014). As in the NCSU (2014) model, students took photos at various library service points using the iPod Touches, and they earned points for each photo. One or two library staff members monitored the photos the students uploaded from the iPods to Evernote and used a Google spreadsheet to tabulate points for competing teams. Another library staff member downloaded the students' photos from Evernote and uploaded them into a PowerPoint presentation. After the scavenger hunt, library staff used the completed PowerPoint presentation as they led the class on a virtual tour of library spaces, which was intended to provide additional information about library services and resources available at the locations pictured. The scavenger hunt and the use of student photos made for a fun and engaging activity. Both students and instructors responded very positively to the scavenger hunt tours, as the verbal and written feedback showed a high level of engagement and the students' photos demonstrated their enjoyment in taking their own photos and viewing pictures from other groups.

Although the iPod Touch scavenger hunt was successful in terms of student engagement, L&O experienced some challenges with this model. The first significant challenge was time. The Academic Integrity sessions were fifty minutes long, and the scavenger hunt portion usually took twenty to thirty minutes. In order to have sufficient time for the virtual tour, the student grades and the PowerPoint presentation needed to be ready as soon as the students finished the scavenger hunt. However, this tight turnaround was difficult because creating the PowerPoint took multiple steps, the grading system was labor-intensive, and the entire process required multiple software applications. Due to staff scheduling limitations, L&O did not always have the preferred number of staff members to score and process the pictures in a timely manner. The second major challenge L&O experienced was hardware limitations. There were only six iPod Touches available, which limited the size of classes to which L&O could offer the scavenger hunt tours. Despite these challenges, L&O learned that self-guided tours and the use of technology were desirable elements for future tours because assessments determined that students found this type of tour to be more engaging than traditional library staff–led tours.

## *Making the Pitch for Technology*

Because the small number of mobile devices was a major limitation for the library's technology-based self-guided tours, L&O knew that it needed more mobile devices. The six first-generation iPod Touch devices were purchased in the late 2000s, upgrading the devices to a recent iOS was becoming increasingly difficult, and many apps and websites wouldn't load properly. It was clear that the iPod Touches needed to be retired or replaced in order to continue exploring mobile technologies for instruction.

L&O took the opportunity to review the current mobile technology choices. Since an upcoming library renovation would demolish several of the library's instruction spaces for an unspecified time, the timing was fortuitous. There was a clear need for mobile technology not only as a tool for active learning, but also as a way to provide temporary instruction spaces during the forthcoming renovation. While librarians enjoyed the portability of the iPod Touch, its screen size limited its utility. Librarians found it difficult to search the catalog, website, or databases, so they were rarely used for typical library instruction purposes. L&O decided that iPad Minis provided the advantage of being portable for use during virtual scavenger hunts, yet they were large enough for searching and navigation during information literacy classroom instruction.

The Director of L&O submitted a proposal to purchase twenty iPad Mini 2 devices. The proposal emphasized using mobile technology to solve a myriad of needs (e.g., pop-up instruction during the renovation, engagement with new technologies, revising standard walking tours, and more). The University Libraries' IT department, Digital Initiatives (DI), helped supply specifications for the devices and ordered them through a pre-existing university contract. As L&O did not anticipate the devices leaving campus Wi-Fi-enabled buildings, it opted to purchase the 16 GB Wi-Fi-only devices. In addition, charging and syncing management was addressed through the purchase of a Parasync i20 docking station. The centrally charged docking station came with twenty specialized cases for the iPad Minis to seat into the dock without individual cables, which made it possible to charge all devices simultaneously. With an added piece of software, L&O was also able to deploy a centralized app load. Finally, DI supplied a rolling cart to house the docking station, iPads, and other technology, such as a mobile projector. Once the initial setup was complete, L&O had a truly mobile classroom.

## *Texas A&M University Gateway Program*

The first test case for the new mobile set-up was a transformation of the traditional library tours provided to the University's Gateway program. The Texas A&M University's Aggie Gateway to Success (Gateway) program is a provisional admission program that enrolls students over the summer to provide them an opportunity to demonstrate their ability to succeed academically in college-level courses (Texas A&M University, 2016a). As part of their participation in the Gateway program, students take the course STLC 289, which focuses on "selected topics in academic development and improvement" (Texas A&M University, 2016b).

For the last few years, each section of STLC 289 has visited Evans Library near the end of the semester in order to familiarize students with the library as an academic resource that can contribute to their success. Prior to 2015, the library provided various versions of in-person, staff-led walking tours for the Gateway classes. By walking to, and through, different service points, students learned how to locate and access the library's collections, services, and other resources, and both students and course instructors provided positive feedback about the library tours. However, Gateway students represent a population of academically at-risk students. For this population, engagement with the help services and resources in the University Libraries can make a dramatic difference in their future academic success. At-risk students can struggle with the traditional classroom experience, typified by lecture-style teaching. Passive library tours were replicating the lecture experience, a pedagogy ill-suited for these students. In an effort to address these concerns, the authors wanted to make the tours more fun, interactive, memorable, and impactful for the Gateway students.

## *Mobile Technology and Gateway Tours*

With some previous experience using mobile technologies for self-guided tours, the authors sought to redesign the Gateway tour. The first step was identifying the library learning outcomes for the tour:

- Students will become more familiar with the Evans Library and Annex physical layout.
- Students will remember basic information about eight key library services and spaces.
- Students will explore a specific library space or service and be able to teach classmates basic information about that topic.
- Students will use mobile technology to incorporate visual elements into a presentation.

The authors created an alignment grid to work backward from the learning objectives to design the session activities (see appendix 4A). After discussing elements that support active learning, the authors determined that in order to encourage student engagement, they wanted to make students responsible for their own learning by having the students lead the tour. Time would not permit students to do a complete self-guided tour, so the tour was designed to send the students in groups to different service points and spaces in the library. Librarians would ask each group to take selfies and answer questions in each area in order to present to the class about the area they visited. By asking students to work in groups and then share their new knowledge and their selfies with the class, the authors intended to foster both group and individual accountability for learning. This accountability was an important element of the selfie tour design because research indicates that group and individual accountability increase the likelihood of fostering a sense of personal responsibility (Johnson & Johnson, 2009). Asking students to present what they learned to their classmates is a powerful way to hold students accountable to each other for their learning. It also gives the librarians a new perspective about the library and what captures the students' attention as they hear what students found unique or helpful about a particular service or space.

In addition to designing the tour activity, the authors needed to develop a strategy for assessing student learning. The Gateway instructors typically incorporate information from the library tour into a quiz later in the semester, but it was also important to develop another assessment method that would gauge learning within the confines of the library tour. Because the tour was designed to engage students in active learning using iPads, the authors determined that the same technology could also be harnessed to capture visual markers of student engagement in the form of selfies. These photos inspired the title of the new tour: the "selfie-guided tour."

## The Selfie-Guided Tour

After L&O received the iPads and confirmed the new lesson plan with the STLC 289 instructors, the authors were ready to implement the new tour. Each class began with one or two librarians introducing themselves and explaining the objectives of the tour. The students then were broken up into seven teams with roughly four people in each team. The librarians distributed and then explained a handout that included (1) instructions for the activity, (2) directions on how to get to the group's designated service point, and (3) three to four prewritten questions the students would have to answer. Librarians also distributed one iPad per group and spent a few minutes logging the students into the iPads. Next, the groups ventured forth to find their assigned service point or space, which included areas such as the library circulation desk, course reserves, quiet study spaces, and more.

When the students arrived at their service point or space, they used the iPad to take a selfie. The students then sought the answers to their questions and were encouraged to seek help from librarians and library staff. The students were also given instructions that if the service point was busy serving other patrons, they were to use the library website or chat service to answer the questions. Once the students had their selfies and answers and had sufficiently explored the space or service point, they returned to the library classroom. Librarians guided them through the process of adding their selfies to a shared iCloud folder so the photos could be presented to the class. Once all groups had returned to the classroom and uploaded their selfies, each group presented their selfies to the class and shared what they had learned about their assigned service or space. Occasionally, librarians supplemented student presentations with clarifying information about a service point or space or additional information that students may have missed. While each group presented, the other students took notes on their classmates' presentations, as this material could be on an upcoming quiz. When the class session was over, the librarians logged the students out of the iPads and returned the iPads to the charging station.

## Assessing the Selfie-Guided Tour

The library had no formal assessment plan for previous versions of the Gateway tours, so there were no direct benchmarks against which to measure the selfie-guided tours. In order to evaluate the impact of the selfie-guided tours, an assessment methodology was implemented that incorporated both formal and informal assessment. Formal assessment was conducted by the Gateway instructors in their classes, in the form of multiple-choice and short-answer quizzes that were formulated by the instructors based upon tour content. Some instructors opted to share quiz data with the librarians. While this data indicated that students had good retention of the information shared during the library sessions, it is not possible to determine whether the results indicate an improved outcome due to the lack of comparison data. However, informal assessment measures were also used in order to determine whether the selfie tours accomplished the objective of increasing student engagement during the library sessions. The selfies taken by students during the tours served as informal assessment artifacts that indicated that students engaged very positively during the selfie-guided tours. The creativity displayed by the Gateway students went far beyond expectations; students posed in highly creative (and sometimes acrobatic) ways and even engaged staff members to participate in their pictures, which led to unanticipated but highly desirable relationship building between students and staff. Many student groups took more pictures than were required and engaged their competitive spirits to have the best selfies. The positive and even boisterous reactions displayed by their classmates during the student presentations indicated that their efforts to create the best pictures were much appreciated.

Although the positive reactions of students were one measure of success, the reactions of instructors and collaborators were also important. Many Gateway instructors reacted positively to the selfie-guided tour sessions and commented on the students' level of engagement. Collaborators within the library also had a positive reaction; for example, a staff member from one department that served as a stop on the selfie-guided tour remarked, "All of the admin staff loved the changes, and the integration into the course. Very smart idea!" However, the tour posed some administrative challenges, especially for Gateway instructors. The instructors had to develop their in-class quizzes based upon the information conveyed in the student presentations, which could vary from one class to another. These types of challenges could be mitigated in future Gateway tours by working with library departments to help them emphasize a consistent message from one group to another and also by simplifying the amount and type of content delivered to students to ensure a consistent message from the library.

## Lessons Learned and Future Directions

Although students, instructors, and librarians all responded positively to the Gateway selfie-guided tour, the authors plan to make changes to future iterations of the tour in order to make them easier for both students and librarians. Mobile technology will continue to play a vital role in the library's student engagement strategy, but refining the approach will help reduce challenges and improve outcomes.

In the initial version of the selfie-guided tour, the authors used iCloud Photo Sharing in order to collect and share student photos. Although this worked fairly well, other apps have since been tested, and it was determined that the app Padlet may work better for the selfie tours because its photo-sharing process is more streamlined. By switching to a new app, the authors hope to minimize the amount of time needed for setup at the beginning of each class as well as the amount of time needed to reset the iPads between classes. This will both maximize the amount of time devoted to learning during the session and provide librarians with extra time to recover, especially between back-to-back classes. The authors are also in the process of revising learning objectives related to the students' familiarity with library services. Although it may be easier to quiz students about library content where there is an easily identifiable correct or incorrect answer, such as the hours a particular library building is open, this type of content is also likely to be the most easily searchable, and therefore not necessarily important to memorize. Instead, the authors are working with representatives from service points around the library to identify the information about their services that they most wish incoming students to know and are developing a plan to convey only the most necessary information to students during the selfie-guided tour.

In addition to refining the selfie-guided tours, L&O has been identifying new uses for the mobile devices. Most recently, the iPads were used as an assessment tool. Library volunteers were stationed at library exits during the library's annual fall Open House event and passed out iPads preloaded with online surveys. This method of gathering feedback was quite effective, and L&O has plans to use the iPads for assessment at future events. Mobile technology will also play an important role during and after the library's renovation project. Once the construction begins in late 2016 and library learning spaces are disrupted, L&O anticipates making heavy use of the mobile technology cart. The typical class size on the Texas A&M campus hovers around forty-six students; however, L&O purchased only twenty iPads as it anticipated that many students would come to an instruction session with their own tablet or laptop device. Mobile devices would be used as a supplement in a BYOD (bring your own device) model during the construction. After construction is complete, the renovated spaces will include both wired library classrooms and flexible BYOD spaces with mobile technology. As L&O gains a better understanding of how these new spaces will impact the library's instruction program, it may look at expanding the number and types of mobile devices.

Additionally, there is a small group of librarians and staff interested in creating an unmediated virtual library tour post-renovation. Prospective students, parents, and alumni often approach the library's front desk asking for individual library tours, which can pose a burden for the library's busy desk staff. However, the planning and creation of a virtual tour is on hold until construction is completed as many of the library's interior spaces will go through dramatic transformation. Until that transformation is complete, stakeholders are actively researching tour technologies and mobile applications that could be used to enhance learning during virtual tours.

## Conclusion

Teaching is a creative endeavor. Good teachers experiment with new ways of exploring their topic, presenting ideas, and engaging in learning activities. However, it is all too easy to rely on previous lesson plans, rehash last semester's PowerPoint, or revert to a traditional walking tour of the library when pressed for time and resources. Introducing technology can serve as a positive disruption, shaking instructors and tour leaders out of these fixed systems. In the case of library tours, introducing carefully selected mobile technologies enabled the authors to flip a historically passive tour of library facilities into an active learning experience. Increasing student engagement through the use of selfie-guided tours recontextualized the library as a fun and welcoming place for the Gateway students, an academically at-risk population. These students in particular warrant special consideration when thinking about library instruction. Due to their status as provisionally admitted students, they may come to the library with increased levels of library anxiety and fear of failure. Using a familiar construct, the selfie, and a familiar tool, the iPad, the authors hope to reduce the risk of failure while increasing active learning.

Introducing technology into library tours has also been a positive experience for library instructors. Interested librarians had the opportunity to collaborate on creating learning outcomes, lesson planning, coteaching, and implementing the technology. The conversations among team members were enthusiastic, fun, and focused on good pedagogical design. The initial experimentation with mobile technologies has led to a series of iterative improvements and experiments with new tools. Furthermore, the project demonstrated the librarians' teaching prowess to the Gateway faculty. They were impressed with the outcomes and are already in discussions about library programming for next year's course.

## APPENDIX 4A. GATEWAY ALIGNMENT GRID

| Learning Objective/ Outcome | How Learning Will be Assessed | Teaching/Learning Activity | Technology Resources |
|---|---|---|---|
| Students will become more familiar with the Evans Library and Annex physical layout. | Students will successfully take photos of themselves in a different place in the library. | Students will divide into groups and each group will be assigned a service or space. Students will navigate their way through the library to find their assigned space/service and will take a selfie of themselves in that space. | iPads configured to take photos and share photos with librarian instructor; Selfie tour instruction handout; Group handout specific to a library service or space. |
| Students will remember basic information about 8 key library services and spaces. | Quiz administered by their instructors. | Students will divide into groups and each group will be assigned a service or space. Groups will answer questions about their service or space and present this information to their classmates. | iPads configured to take photos and share photos with librarian instructor; Selfie tour instruction handout; Group handout specific to a library service or space. |
| Students will explore a specific library space or service and be able to teach classmates basic information about that topic. | Students will present to their classmates about a library service or space and will provide answers to all worksheet questions on their topic. | Students will divide into groups and each group will be assigned a service or space. Groups will answer questions about their service or space and present this information to their classmates. | iPads configured to take photos and share photos with librarian instructor; Selfie tour instruction handout; Group handout specific to a library service or space. |

| Learning Objective/ Outcome | How Learning Will be Assessed | Teaching/Learning Activity | Technology Resources |
|---|---|---|---|
| Students will use mobile technology to incorporate visual elements into a presentation. | Students will use their selfies in their presentation and will refer to the images when teaching about the space. | Students will divide into groups and each group will be assigned a service or space. Students will take a selfie(s) of themselves in that space and will use this selfie in their presentation. | iPads configured to take photos and share photos with librarian instructor; Selfie tour instruction handout; Group handout specific to a library service or space. |

## References

Bonwell, C. C. & Eison, J. A. (1991). *Active learning: Creating excitement in the classroom.* Washington, D.C.: George Washington University School of Education and Human Development. Retrieved from http://files.eric.ed.gov/fulltext/ED336049.pdf.

Burke, A. (2012, July 30). Demystifying the library with game-based mobile learning [Blog post]. Retrieved from http://acrl.ala.org/techconnect/post/demystifying-the-library-with-game-based-mobile-learning.

Burkhardt, A. & Cohen, S. F. (2012). "Turn your cell phones on": Mobile phone polling as a tool for teaching information literacy. *Communications in Information Literacy, 6*(2), 191–201. Retrieved from http://www.comminfolit.org/.

Cairns, V., & Dean, T. C. (2009). Creating a library orientation for the iPod generation. *Tennessee Libraries, 59*(2). Retrieved from http://www.tnla.org/?300.

Calkins, K., & Bowles-Terry, M. (2013). Mixed methods, mixed results: A study of engagement among students using iPads in library instruction. In *Imagine, Innovate, Inspire: The Proceedings of the ACRL 2013 National Conference* (pp.423–428). Chicago: Association of College & Research Libraries. Retrieved from http://www.ala.org/acrl/sites/ala.org.acrl/files/content/conferences/confsandpreconfs/2013/papers/Calkins_Mixed.pdf.

Cook, D., & Sittler, R. (Eds.). (2008). *Practical pedagogy for library instructors: 17 innovative strategies to improve student learning.* Chicago: Association of College & Research Libraries.

Detlor, B., Booker, L., Serenko, A., & Julien, H. (2012). Student perceptions of information literacy instruction: The importance of active learning. *Education for Information, 29*(2), 147–161. doi:10.3233/EFI-2012-0924.

Drueke, J. (1992). Active learning in the university library instruction classroom. *Research Strategies, 10*(2), 77–83.

Fawley, N. & Krysak, N. (Eds.). (2016). *Discovery tool cookbook: Recipes for successful lesson plans.* Chicago: Association of College and Research Libraries.

Foley, M. & Bertel, K. (2015). Hands-on instruction: The iPad self-guided library tour. *Reference Services Review, 43*(2), 309–318. doi:10.1108/RSR-07-2014-0021.

Freeman, S., Eddy, S.L., McDonough, M., Smith, M. K., Okoroafor, N., Jordt, H., & Wenderoth, M. P. (2014). Active learning increases student performance in science, engineering, and mathematics. *Proceedings of the National Academy of Sciences, 111*(23), 8410–8415. doi:10.1073/pnas.1319030111.

Gibeault, M.J. (2015). Using iPads to facilitate library instruction sessions in a SCALE-UP classroom. *College & Undergraduate Libraries, 22*(2), 209–223. doi:10.1080/10691316.2014.924844.

Havelka, S. (2013). Mobile information literacy: Supporting students' research and information needs in a mobile world. *Internet Reference Services Quarterly, 18*(3–4), 189–209. doi:10.1080/10875301.2013.856366.

Hoppenfeld, J. (2012). Keeping students engaged with web-based polling in the library instruction session. *Library Hi Tech, 30*(2), 235–252. doi:10.1108/07378831211239933.

Ingalls, D. (2015). Virtual tours, videos, and zombies: The changing face of academic library orientation. *Canadian Journal of Information and Library Science, 39*(1), 79–90. doi:10.1353/ils.2015.0003.

Johnson, D. W., & Johnson, R. T. (2009). An educational psychology success story: Social interdependence theory and cooperative learning. *Educational Researcher, 38*(5), 365–379. doi:10.3102/0013189X09339057.

Kearns, A. (2010). An iPod (MP3) library tour for first-year students. *College & Undergraduate Libraries, 17*(4), 386–397. doi:10.1080/10691316.2010.525427.

LevelUp. (n.d.). Retrieved from https://www.thelevelup.com/.

Lippincott, J. K. (2010). A mobile future for academic libraries. *Reference Services Review, 38*(2), 205–213. doi:10.1108/00907321011044981.

Maloney, M. M., & Wells, V. A. (2012). iPads to enhance user engagement during reference transactions. *Library Technology Reports, 48*(8). 11–16. Retrieved from https://journals.ala.org/ltr/article/download/4288/4919.

Marcus, S., & Beck, S. (2003). A library adventure: Comparing a treasure hunt with a traditional freshman orientation tour. *College & Research Libraries, 64*(1), 23–44. doi:10.5860/crl.64.1.23.

Mawson, M. (2007). iPod tours: A new approach to induction. *New Review of Information Networking, 13*(2), 113–118. doi:10.1080/13614570801900021.

McMunn-Tetangco, E. (2013, April). If you build it…? One campus' firsthand account of gamification in the academic library. *College & Research Libraries News, 74*(4), 208–10. Retrieved from http://crln.acrl.org/content/74/4/208.

Michael, J. (2006). Where's the evidence that active learning works? *Advances in Physiology Education, 30*(4), 159–167. doi:10.1152/advan.00053.2006.

Mikkelsen, S., & Davidson, S. (2011). Inside the iPod, outside the classroom. *Reference Services Review, 39*(1), 66–80. doi:10.1108/00907321111108123.

Miller, K., & Putnam, L. (2015). More of the same? Understanding transformation in tablet-based academic library instruction. *Internet Reference Services Quarterly, 20*(3–4), 105–126. doi:10.1080/10875301.2015.1092188.

National Institute of Education (U.S.). (1984). *Involvement in learning: Realizing the potential of American higher education: Final report of the Study Group on the Conditions of Excellence in American Higher Education.* Washington, D.C.: National Institute of Education, U.S. Department of Education. Retrieved from http://hdl.handle.net/2027/mdp.39015021483196.

North Carolina State University Libraries. (2014, June 13). *Instruction support services: NCSU Libraries mobile scavenger hunt*. Retrieved from https://www.lib.ncsu.edu/instruction/scavenger-details.php.

Oling, L., & Mach, M. (2002). Tour trends in academic ARL libraries. *College & Research Libraries, 63*(1), 13–23. doi:10.5860/crl.63.1.13.

Pagowsky, N. (2013, May 13). Taking a trek with SCVNGR: Developing asynchronous, mobile orientations and instruction for campus [Blog post]. Retrieved from http://acrl.ala.org/techconnect/post/taking-a-trek-with-scvngr-developing-asynchronous-mobile-orientations-and-instruction-for-campus.

Prince, M. (2004). Does active learning work? A review of the research. *Journal of Engineering Education, 93*(3), 223–231. doi:10.1002/j.2168-9830.2004.tb00809.x.

Sandy, J. H., Krishnamurthy, M., & Rau, W. (2009). An innovative approach for creating a self-guided video tour in an academic library. *Southeastern Librarian, 57*(3), 29–39. Retrieved from http://digitalcommons.kennesaw.edu/seln/vol57/iss3/5.

Sciammarella, S., & Fernandes, M. I. (2007). Getting back to basics: A student library orientation tour. *Community & Junior College Libraries, 14*(2), 89–101. doi:10.1300/02763910802139157.

Sittler, R. L. & Cook, D. (Eds.). (2009). *The library instruction cookbook*. Chicago: Association of College & Research Libraries.

Texas A&M University. (2016a). *Aggie gateway to success*. Retrieved from https://tap.tamu.edu/gateway/.

Texas A&M University. (2016b). *STLC- student learning center*. Retrieved from http://catalog.tamu.edu/undergraduate/course-descriptions/stlc/.

Texas A&M University Libraries. (2014). *Academic integrity development program*. Retrieved from http://guides.library.tamu.edu/academicintegrity.

Virginia Tech University Library. (n.d.). *Newman library self-guided tour*. Retrieved from http://m.lib.vt.edu/tour.php.

Whitchurch, M. J. (2015). *Library tour evolution: Analog, digital, mobile*. Retrieved from http://scholarsarchive.byu.edu/facpub/1546.

CHAPTER 5*

# Beyond Passive Learning:
## Utilizing Active Learning Tools for Engagement, Reflection, and Creation

*Teresa E. Maceira and Danitta A. Wong*

## Introduction

The twenty-first-century student frequently engages with mobile environments to fulfill his or her information needs. Reports from the Pew Research Center (Duggan, 2015) indicate an increasing trend in the use of mobile social media platforms in the US adult population. According to the report *Social Media Usage: 2005–2015 (2015)*, 65 percent of American adults use social networking sites, and young adults (ages 18–29) reported the highest social media usage for all age groups, at 90 percent. Among the young adult age group, social media usage in 2005 was reported to be 12 percent (Perrin, 2015). This sharp 650 percent rise illustrates the increasingly networked environment inhabited by young adults. In an effort to engage students in the online environment they frequent, educators at the University of Massachusetts Boston utilized iPads to integrate mobile technology into library instruction and other teaching initiatives. The impetus for exploring emerging technologies for library instruction was par-

---

* This work is licensed under a Creative Commons Attribution 4.0 License, CC BY (https://creativecommons.org/licenses/by/4.0/).

ticipation in University of Massachusetts Boston's iPads in the Classroom program. As part of this program, the instruction librarians acquired an iPad cart for library instruction in fall 2014 in order to facilitate and create increased and enhanced teaching opportunities. The authors wanted to go beyond substituting the iPads for desktop computers to transforming tasks so that learners could engage in higher order skills of creating and evaluating information, as defined by Puentedura's Substitution Augmentation Modification Redefinition (SAMR) model of technology integration (2014) and incorporate best practices for the use of technology to promote learning.

This chapter will discuss activities that integrate iPads into library instruction and highlight the use of research guides, web-based polling, gaming pedagogy, online surveys, and other web-based applications for academic research that participants directly engage with, reflect upon, and use to create information in transformative ways. Both in course-specific information literacy sessions with students and iPad workshops focusing on educational applications open to the University of Massachusetts Boston community will be discussed. The aim is to address questions such as these: Does integrating iPads into library instruction sessions enhance the experience by adding value or detract by creating obstacles? To what extent are we integrating the technology into our sessions? Is this technology just another tool? How do apps add value to learning? With any teaching innovation, there are lessons to learn. This chapter will discuss the benefits and drawbacks of teaching with technology and provide suggestions on how to integrate emerging technologies into the ever-evolving classroom.

## Literature Review
### *Metaliteracy*

Mackey and Jacobson (2011) first popularized the term *metaliteracy* to incorporate changing technologies and different literacy types into library instruction. Acknowledging the centrality of technology, the Association of College and Research Libraries (ACRL) *Framework for Information Literacy for Higher Education* draws upon the concept of metaliteracy in its discussion of the evolving digital environments that impact our understanding of information literacy (Association of College and Research Libraries [ACRL], 2016). Thus the ACRL framework and the concept of metaliteracy provide a foundation for integrating technology with information literacy by recognizing the dynamic and collaborative elements of information creation in a participatory networked environment. Mackey and Jacobson (2011) maintain that metaliterate learners are consumers, producers, and sharers of online information in collaborative spaces, stressing the primary importance of information, whereas the medium is secondary.

## *Substitution Augmentation Modification Redefinition (SAMR)*

The authors used concepts from metaliteracy and the SAMR model to effect a more purposeful and reflective use of technology in the classroom. Puentedura (2014) developed the SAMR model to guide educators with the design, implementation, and assessment of teaching that integrates technology to transform tasks and target increasingly complex learning outcomes. In the initial substitution stage of SAMR, the new technology serves as a direct substitute without a functional change of the task. In the second stage of augmentation, the new technology provides some functional improvement. The first two stages of the model allow for some enhancement of educational tasks and typically target the learning outcomes of remembering, understanding, and applying information. An example of substitution is to have learners do a keyword search for articles using a database app on an iPad rather than on a computer. This involves having learners apply and understand search techniques to find articles much in the same way that it is done on a computer. The last two stages of the model, modification and redefinition, allow for transformation of tasks and the application of more cognitively complex processes of analysis, evaluation, and creation as outlined in the revised Bloom's taxonomy (Heer, 2009). An example of redefinition would be to have learners create and share a video presentation using a screen casting tool such as Educreations instead of giving a presentation in the classroom. The creation of a digital product transforms the task of presenting through the incorporation of an online participatory environment. In this way, the authors used the SAMR model to be selective and goal-focused in using technology to teach, as recommended by Miller (2014). In short, the lesson outcomes incorporated varying levels of technology-integrated tasks so that learners could sequentially develop increasing comfort with the technology and use it to eventually perform more sophisticated cognitive processes such as the creation of products that could be shared online (see table 5.1).

## *Mobile Information Literacy*

The research literature reflects the increasing use and benefits of mobile technology in libraries. Havelka (2013) observed emerging technologies, and in particular mobile technology integration, in many academic library services with the exception of information literacy sessions. Havelka (2013) inferred that students would welcome mobile information literacy instruction because surveys showed that students would consider using iPads as their only research tool for academic purposes. Fabian and MacLean (2014) reported that the use of mobile devices fostered student engagement and collaboration. Furthermore, mobile environments

**TABLE 5.1**
Tools and activities using technology incorporating the SAMR model.

| Functional Category | Tool | Activity | Activity Description | Learning Outcomes | SAMR Model |
|---|---|---|---|---|---|
| Presentation & Collaboration | Padlet | Knowledge sharing | Crowdsource the course topic | Evaluating and using appropriate resources for research<br><br>Contribute to class knowledge | Modification |
| | Haiku Deck | Presentation | Create of a product | Utilizing technology to express an idea | Modification |
| | Educreations | Video creation | Present on the paper topic and appropriate resources | Evaluating and using appropriate resources for research | Modification |
| | Poll Everywhere | Knowledge sharing | Crowdsource the course topic | Evaluating and using appropriate resources for research<br><br>Contribute to class knowledge | Modification |
| Feedback, Reflection, Assessment | Research Guide Poll | Voting | Vote on preferred search tool used in the workshop to initiate a discussion of tool features | Evaluating and using appropriate resources for research | Augmentation |

**TABLE 5.1**
Tools and activities using technology incorporating the SAMR model.

| Functional Category | Tool | Activity | Activity Description | Learning Outcomes | SAMR Model |
|---|---|---|---|---|---|
| Feedback, Reflection, Assessment | Padlet | Brainstorming | Brainstorm and suggest search strategies | Analyze the credibility of search results<br><br>Refine search strategies | Augmentation |
| | Answer Garden | Reflection assessment | Identify and share credible resources | Evaluating credible resources | Augmentation |
| | Kahoot | Assessment game | Answer reflective survey questions and demonstrate understanding of concepts | Evaluating credible resources | Redefinition |
| | Socrative | Reflection assessment | Answer reflective survey questions | Demonstrate self-reflection & understanding of material | Augmentation |
| Citation | EasyBib | Citation | Generating citations | Attribution | Augmentation |
| | RefME | Citation | Generating citations | Attribution | Augmentation |

made lessons meaningful, while the apps utilized added value to the instruction. In their experience, the mobile nature of the devices facilitated improved interactions between students and faculty by reconfiguring the physical space and introducing innovative app-specific activities.

## Mobile Applications and Librarians

In response to the increasing academic research conducted in mobile environments, the authors developed a series of workshops focusing on apps. The rise in the use of mobile devices has generated an upsurge of associated apps. According to the website Statista (https://www.statista.com), from June 2008 to June 2016 the number of Apple apps grew to 2 million ("Number of available apps", 2016). Recognizing the increasing use of mobile devices and apps, Havelka and Verbovetskaya (2012) make the argument that mobile information literacy is a necessary skill that librarians should introduce into information literacy classes. Spina (2014) states that librarians are well placed to help library users to navigate this constantly evolving environment. Similarly, Hennig (2014) states that with the proliferation of apps, it's incumbent on librarians to be app-literate. Hennig (2014) further reaffirms that librarians need to become more knowledgeable about apps to impart knowledge and create teaching opportunities. Although student interest in utilizing mobile devices for education is evident, the majority lack sufficient skills in evaluating apps and mobile websites (Havelka, 2012; Yarmey, 2011). Canuel and Crichton (2015) and Hennig (2014) state that by providing workshops to disseminate information on apps, librarians address the academic needs of their students. Spina (2014) and Hennig (2014) outline criteria to use when evaluating apps and for sharing the information through various methods.

# Tools and Learning Activities

The technology integration in information literacy classes at the University of Massachusetts Boston relied mostly on free web-based tools and apps. The authors actively integrated Padlet, AnswerGarden, and Poll Everywhere. The flexibility of these tools made them perfect for a wide variety of activities such as polling, reflection, and assessment. Library subscription–supported apps, such as BrowZine, FT (from *Financial Times*), and LibGuides, were also integrated into the classes for citing, presenting, researching, and collaborating in the information literacy classes, mobile sessions, and workshops. Learning activities included a description of the tools to be utilized and the applicable level of technology integration as defined in the SAMR model, in addition to identified goals for each session.

## *Presentation, Collaboration, and Assessment Tools*
### EDUCREATIONS
The Educreations app is a recordable whiteboard that facilitates the creation of short videos. Educreations users can create multiple interactive whiteboards; import videos, links, and images; and share videos with other Educreation users. Educators utilize Educreation videos as vehicles to prompt students to explain a topic or an idea.

Four freshman English (ENGL 101) classes had mobile information literacy sessions utilizing iPads, a research guide, and the Educreations app. In the information literacy session, facilitators measured student knowledge and shifted away from using the iPad as a substitute for computers and moved to the transformative level as defined in the SAMR model. Instead of students searching for information for their papers and e-mailing or creating a Word document on their sources, they created and shared videos documenting their research ideas.

### *Learning Outcomes*
The instruction sessions intended learning outcomes included the following:
1. Identifying key concepts and terms related to a research question
2. Applying keyword and Boolean search techniques
3. Evaluating and using appropriate resources for research

The ENG101 course was part of the iPads in the Classroom program, and therefore the students used iPads weekly and the sessions were designed to integrate information literacy into a mobile environment. Facilitators guided students on the iPads in conducting database searches for scholarly material, identifying relevant keywords and subject terms. As the sessions progressed, students engaged in a continuous dialogue with the librarian and professor, who functioned as co-facilitators and co-learners. Facilitators posed leading questions, such as: "How do I know this is a research article not a literature review?" "What are the differences between conducting a keyword search versus a subject search?" and "How would you cite a YouTube comment?" The collaborative research taking place in the classroom created a social learning environment by changing the students' role from passive receivers of information to active partners in the discovery process utilizing iPads as the medium.

In the second half of the sessions, the students used Educreations to create videos on their chosen topic. The aim of the video creation exercise was to reinforce information literacy skills by conducting academic research, creating a resources list to use for the assignment, and employing an effective and fun tool. The students' Educreations recordings included sketches describing public spaces, images retrieved through Google Images, imported citations from the databases, and website links. According to the SAMR model, the four classes progressed from the substitution stage with the deployment of the iPads as computer substitutes to the redefinition stage by incorporating the Educreations app because it allows students to create a product that was not possible without the technology.

### *Challenges*
Technical problems such as occasionally spotty Wi-Fi connectivity and limitations of the iPads' microphones hampered the quality of the recordings. Further technical issues arose from the students' lack of experience using the Educreations app and time constraints. Initially, the time allotted for the exercise was twenty-five minutes in a seventy-five-minute class session, including a demonstration on utilizing the app and exercise instructions. Ideally, facilitators should have allocated additional time to explore and practice using the app.

### *Positive Outcomes*
Classroom dynamics changed with the adoption of the iPads into the information literacy sessions. In each case, the interactions between the facilitators and students during the sessions were fluid and spontaneous, while conversations resulted from individual observations and questions. Fabian and MacLean (2014) noted the collaborative aspect and the "seamless workspace" fostered by the mobile environment. For example, during one of the sessions, a student asked how to cite images found on Google. This question resulted in the class searching Google Images, and a discussion about Google's usage rights options developed. The spontaneity of the mobile environment and the sharing of information and ideas cultivated a collaborative atmosphere. The critical evaluation of information in this dynamic environment aligns with the core fundamentals of metaliteracy, where students act as active searchers and evaluators of information. The process of participating in a conversation involving experts and novices reinforced learning concepts where students became active participants in the evaluation of the information sources for validity and reliability.

## PADLET/POLL EVERYWHERE
Padlet is an interactive web-based bulletin board with a variety of uses. Padlet can be used online, embedded into a research guide, shared via social media, or e-mailed. Padlet is ideal for collaboration and posing open-ended questions for reflection and assessment.

The authors used Padlet for identifying source types and sharing search strategies in a workshop for a science seminar. The Padlet in the course guide asked the students to try a web search with specific terms and then to create and post strategies on the Padlet that could improve the quality of results. The class then tried the search strategy recommendations together. This activity reinforced the idea that searching is iterative and allowed the students an opportunity to evaluate strategies based on the results. This activity falls into the augmentation stage of SAMR because having students post suggestions online is a functional improvement over oral responses.

## *Learning Outcomes*
The instruction sessions intended learning outcomes included these:
1. Analyze the credibility of search results.
2. Refine search strategies.

Another Padlet activity for an interdisciplinary seminar workshop included instructor-provided images and text describing potential sources of varying types relevant to the class assignment on wrongful legal convictions (figure 5.1). Each group of students was assigned a source on the Padlet wall and determined if the source was primary or secondary. Then they moved that source to a designated area for either primary or secondary sources on the Padlet wall. Subsequently, they created a citation for the source to add to the Padlet. Here the authors apply the modification stage of SAMR because the technology allows the students to evaluate information and complete tasks without a nontechnical equivalent.

**FIGURE 5.1**
Screenshot of seminar Padlet.

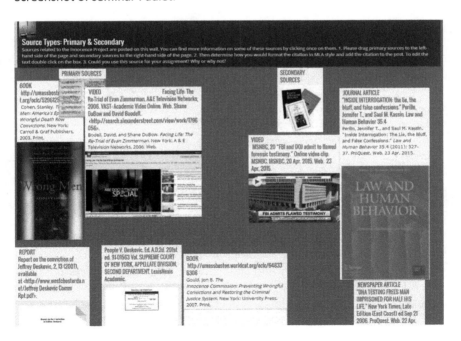

The freshman English (ENGL 102) classes utilized Poll Everywhere and Padlet in the first visit. Poll Everywhere is a web-based polling tool applied in this particular instance as a collaboration tool. The classes created a class bibliography developed through a crowdsourcing activity focused on the concepts surrounding the individual and society. The integration of Padlet and Poll Everywhere into the information literacy session created a collaborative space via the research guide where all four classes contributed to a single platform. This is an example of applying emerging technologies in collaborative spaces, as discussed by Mackey and Jacobson (2014).

### *Learning Outcomes*
The learning outcomes included these:
1. Evaluate and use appropriate resources for research.
2. Apply keyword and Boolean search techniques.
3. Contribute to class learning and utilize technology to express an idea.
4. Identify the differences between paraphrasing and quoting in MLA.
5. Generate citations.

The sessions addressed information literacy subjects in a mobile environment where each attendee (student, the professor, and the librarian) had iPads. The students posted resources including books, book chapters, articles, and websites. Subsequently, students experiencing technical problems with the tools or the iPads went so far as e-mailing their contributions to the librarian for inclusion in the class bibliography. Consequently, the crowdsourcing exercise developed into an informal assessment gauging student comprehension of concepts discussed during the class sessions.

The use of the iPads and the web-based tools for collaboration created a collegial atmosphere with opportunities to explore learning among class peers in association with the professor and the librarian. The crowdsourcing exercise reflected the heightened level of participation and enthusiasm the students exhibited over the opportunity to share knowledge. Johnston and Marsh (2014) also observed that active participation with technology in information literacy exercises promoted higher student engagement by fostering student collaboration. This is another example of students as creators of information in collaborative environments as outlined by Mackey and Jacobson (2014).

Havelka (2013), Yarmey (2011), and Fabian and MacLean (2014) observed higher levels of engagement and social learning exhibited by students in library instruction sessions incorporating mobile technology. Fabian and MacLean (2014) go further to speculate that the novelty of using a device could have added to the demonstrated enthusiasm, although all agree that technology enabled the students to accomplish tasks that would normally not be feasible without technology integration.

The utilization of the crowdsourcing exercise as an informal assessment to measure student understanding of acceptable sources uncovered students' reflection of the differences between scholarly and popular sources. The prompt asked students to contribute to the class knowledge by including resources they personally found useful or interesting. The majority of students contributed resources from the databases even though they didn't have to. The crowdsourcing activity focused on sharing information through group interactions by disseminating group knowledge through a virtual environment. The four ENGL 102 sections contributed to one guide, making the information available to all users of the class guide and beyond to a wider online audience.

## HAIKU DECK

In the second visits to the freshman English (ENGL 102) classes, the activity focused on the creation of a product for the final assignment. The app Haiku Deck provides a simplified process for creation of slides and the incorporation of images to create a visually appealing and impactful presentation. The activity asked students to introduce the rest of the class to their topic by creating three slides on Haiku Deck. Instead of submitting a written proposal on their topics, the students created a visual presentation.

The presentations in the second information literacy sessions included the sophisticated use of images to represent the students' research topics. Students exhibited a high level of interest and personal investment in representing their topics visually. One student, for example, imported his personal images to Haiku Deck to create his own deck containing six slides. However, it was evident that other students lacked comfort with technology and did not complete the activity.

The emphasis of metaliteracy on the production and contribution to the scholarly conversation influenced the development of the lesson plans for these classes (ACRL, 2016) as did the ideas of active engagement reinforced throughout metaliteracy; the notion of students as contributors to knowledge, not as passive consumers; and the realization that information creation can take place in different formats and environments (Mackey & Jacobson, 2014). Producing a product—an Educreations video, or a Haiku Deck presentation—involved a discussion of the value and purpose of the tools for content creation. Essential for the information-literate student is an understanding of differing formats and when to use a particular tool for online creation and collaboration (Mackey & Jacobson, 2011). Correspondingly, as stated by Mackey and Jacobson (2014), the information-literate individual needs to develop the understanding and awareness of the impact and layers of knowledge creation.

The incorporation of iPads in the ENGL 102 classes provided the motivation to use the iPads not as a tool for substitution, as identified in the SAMR Model (Puentedura, 2014), but to move toward the transformative level of modification and redefinition, where the use of iPads promotes student collaborations and social learning.

## ANSWERGARDEN

AnswerGarden is a web-based feedback tool. The authors used the tool to solicit feedback to reflective questions and assessment. The tool AnswerGarden provided assessments that identified knowledge gaps or concepts that needed further reinforcement from the librarians. Questions such as "How do you start your research?" or "Identify primary versus secondary sources." provided instructional opportunities for the librarian to follow up on a misconception or to clarify a concept.

## SOCRATIVE/KAHOOT/LIBGUIDE POLL

The authors used LibGuide surveys and the free assessment platforms Socrative and Kahoot! to administer assessment and reflective survey questions to learners. Miller (2014) points out that the wide availability of online quizzing tools allows instructors to take advantage of the testing effect. The testing effect, reported in numerous studies, finds that testing strongly promotes memory of material (Miller, 2014). Socrative is particularly optimal for formative assessment because it allows the instructor to pose extemporaneous questions. Additionally, online technology allows for immediate autograding and rapid feedback with explanations of answers. Information literacy instructors can create their own tests or reuse free quizzes found in MERLOT, Kahoot!, or other open educational resources.

The authors used Socrative to have students answer questions individually and display answers anonymously as discussion starters, or the answers were used as feedback for instructors to see what the students understood. In a graduate chemistry workshop, students answered reflective survey questions via Socrative such as "Reflect on the different databases and/or tools you have explored in the workshops (Web of Science, Google Scholar, citation managers, bibliometric tools, and social networking tools). Identify one and describe how it could be beneficial in your research or studies." Much in the same way, the LibGuide poll was used to solicit student feedback on favorite tools (see figure 5.2).

Kahoot! has an added advantage of maximizing motivation through the use of game-like elements. Miller (2014) identified many of these elements, including multiple sources of feedback, such as music, sound effects, and points. With Kahoot! you can create a quiz or choose from a variety of freely available prewritten quizzes. The quizzes are played in a group setting. Players answer on their own devices, while the quiz questions are displayed on a projected shared display. Throughout the quiz, players receive points for answering quickly, and the names of the top scorers are displayed in a leaderboard. The authors also used LibGuide polls to stimulate reflection and to assess the students' understanding of tool features. These assessment tools represent the modification stage of SAMR, as these activities could be done orally. However, the autograding, rapid feedback, and game-like features serve to redesign these tasks.

**FIGURE 5.2**
Screenshot of poll.

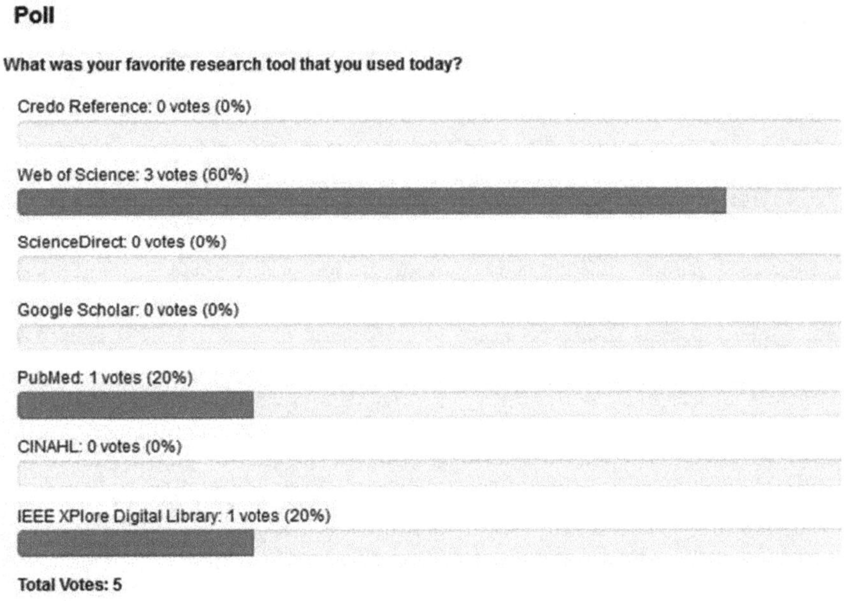

**Poll**

What was your favorite research tool that you used today?

Credo Reference: 0 votes (0%)

Web of Science: 3 votes (60%)

ScienceDirect: 0 votes (0%)

Google Scholar: 0 votes (0%)

PubMed: 1 votes (20%)

CINAHL: 0 votes (0%)

IEEE XPlore Digital Library: 1 votes (20%)

Total Votes: 5

## *Citation Tools—EasyBib/RefME*

Mackey and Jacobson (2014) indicate that in digital environments, attribution can be confusing and challenging; thus, the information-literate student needs to understand the shifting environment and how to cite correctly. Discussions on ethical attribution occurred when using Haiku Deck images and when searching Google Images to include in Educreations videos. Practical application of citation-generating apps EasyBib and RefME in the information literacy classes created learning opportunities on ethical attribution.

# Apps Workshops

The objective of the app workshops is to communicate the added value that specialized program applications (apps) bring as tools for academic research. The literature on mobile technology advocates for librarians taking an active role in imparting knowledge associated with apps (Hennig, 2014; Canuel & Crichton, 2015; Havelka & Verbovetskaya, 2012). The ubiquity of apps in present-day society spans all aspects of an individual's life. Apps used in social interactions, reading, shopping, travel, and business grows more pronounced every day, yet educational applications were not widely known by our students. A search in iTunes for education-related apps reveals a bewildering list of apps, offering little guidance on the app relevance. The goals of the workshops are to augment the academic abilities of the information-literate individual through the enhancement of app literacy.

The designated apps for the workshops address specific aspects of scholarly research. The applicable educational categories targeted included conducting research, file sharing, productivity, accessibility, citing, collaboration, and presentation. The criteria for identifying apps with educational applications included free apps or apps obtained through database subscriptions, apps available in multiple platforms, and ease of use.

Students, faculty, and staff who attended the workshops expressed an increased awareness of education apps and furthermore affirmed that going forward, education apps would be a part of their research skill set. This survey remark exemplified the typical feedback received: "I didn't know there were apps out there that can help me with my research." The apps workshops continually evolve; therefore, changes in the rotation of the featured apps is ongoing. The development and implementation of the workshops address the knowledge gap regarding mobile educational technology. Librarians are well positioned to evaluate and introduce apps that have educational functionality.

The mobile information literacy sessions featuring mobile sites and apps provided the impetus for the development of further outreach in mobile instruction. Canuel and Crichton (2015) observed the increased merging of mobile technology into information literacy classes. The mobile information literacy sessions feature subject-specific apps for business and nursing classes featuring the Financial Times app FT, the Census Bureau economic indicators app, the US National Library of Medicine app PubMed for Handhelds (PubMed4Hh), citation apps EasyBib and RefME, the EBSCOhost app, and the Gale database app AccessMyLibrary. The mobile workshops generated a proactive integration of apps into information literacy sessions. The workshops provide another venue to promote and enhance the academic skill set of students, faculty, and staff through the dissemination of apps with educational applications.

## Technical Issues

The challenges in integrating technology into the authors' information literacy classes echo similar observations made by Havelka (2013) and by Fabian and MacLean (2014). Wi-Fi connectivity, browser issues, and database functionality were the biggest stumbling blocks. Miller (2014) recommends having a contingency plan in case the technology fails. One example of such a plan would be to use nontechnical tools such as paper for conducting a survey or have a spoken discussion.

Wi-Fi proved problematic, especially for mobile instruction sessions outside of the library. Bandwidth could also be a problem in the library instruction room when many individuals log in to some of the web-based tools such as Kahoot! or AnswerGarden. Technical proficiency of the attendees at times presented problems. Information literacy classes and the workshops could be derailed by the participants' comfort level with devices and technology. It must be noted that given a choice, a marked number of students opted for using their laptops because of connectivity problems and lack of full functionality found in mobile applications. Canuel and Crichton (2015) also commented on this issue regarding functionality: while the mobile searching experience is beneficial for short-term research, performing rigorous research on an iPad has the potential to become a frustrating experience.

Database apps such as EBSCOhost and Gale required authentication. The process of authenticating an app is a disruption to an information literacy session. IT authenticates the apps prior to the session to remediate this problem. The survey results from the workshops indicated that users preferred apps that could be directly and immediately employed. The need for accounts for some of the apps presented an obstacle, hence the creation of library e-mail accounts for this purpose. Created accounts made accessing the technology a seamless process in information literacy classes and workshops.

Issues with the internal library website and database functionality presented themselves as the authors moved more of the instruction onto the iPads. For example, the library database A–Z list did not work on the iPads. Databases lacking mobile websites do not display well in iPads, which is problematic.

The time constraints experienced in information literacy sessions factored into the use of technology. The implementation of a tool and its effectiveness can be compromised in fifty- to seventy-five-minute classes. One-shot sessions required careful time allocation because of the added elements of distributing and collecting iPads and providing instruction on the tools and tasks. Keeping up with the literature on education tools is a constant challenge. Tools evolve and features change, which makes it necessary to continually keep up-to-date with the literature. Furthermore, web-based tools and apps require testing and evaluation.

## Conclusion

The acquisition of an iPad cart provided the motivation to move toward further integrating technology into information literacy classes and the opportunity to launch app workshops in the library. Prior to acquiring the cart, the authors used technology in a fragmented manner. The growing pedagogical literature on the overlap between mobile technology and information literacy inspired the authors to actively use these tools (iPads, research guides, web-based tools, and apps), to empower the information-literate individual to produce information. The authors noted the beneficial outcomes in integrating technology into information literacy classes with regard to the level of engagement, creativity, and reflection from the participants in active learning scenarios. First, the tools engage students in active learning tasks so that more time is spent practicing skills than passively receiving information in lectures. Like Johnston and Marsh (2014) and Havelka (2013), the authors reported enthusiastic responses from students towards technology integration in information literacy classes. Being mobile impacted the level of engagement, as Havelka (2013) observed that the realignment of the physical space while using an iPad allowed for more face-to-face interactions among students, faculty, and librarians while they shared devices in a collaborative and synchronous environment. Moreover, reticent students, who would normally be slow or unresponsive to oral queries from librarians, had an opportunity to engage through written responses using technology. Secondly, the authors observed evidence of creativity in the students' products and in the thoughtful integration of different applications in information literacy classes. Finally, the benefit of increased reflection resulted from the process of students responding to questions via AnswerGarden, Socrative, and Padlet in parallel to the participatory digital environments in their daily lives. Using these tools, the students read, created, and commented on other students' responses about the classroom concepts and activities in the same way that they regularly interact with user-generated content in the form of *Wikipedia*, Facebook, Twitter, and comments at the end of online articles. Overall, the benefits of increased engagement, creativity, and reflection outweighed the technical drawbacks of integrating mobile technologies in library instruction.

# References

Association of College and Research Libraries. (2016, January 11). Framework for Information Literacy for Higher Education. Retrieved from http://www.ala.org/acrl/standards/ilframework.

Canuel, R., & Crichton, C. (2015). Leveraging apps for research and learning: A survey of Canadian academic libraries. *Library Hi Tech*, 33(1), 2–14. http://doi.org/10.1108/LHT-12-2014-0115.

Duggan, M. (2015, August 19). *Mobile messaging and social media 2015*. Retrieved from http://www.pewinternet.org/2015/08/19/mobile-messaging-and-social-media-2015/.

Fabian, K., & Maclean, D. (2014). Keep taking the tablets? Assessing the use of tablet devices in learning and teaching activities in the Further Education sector. *Research in Learning Technology*, 22. https://doi.org/http://dx.doi.org/10.3402/rlt.v22.22648.

Heer, R. (2009, March). A model of learning objectives based on a taxonomy for learning, teaching, and assessing: A revision of Bloom's Taxonomy of educational objectives. Retrieved from http://www.celt.iastate.edu/teaching/effective-teaching-practices/revised-blooms-taxonomy.

Havelka, S. (2013). Mobile information literacy: Supporting students' research and information needs in a mobile world. *Internet Reference Services Quarterly*, 18(3–4), 189–209. http://doi.org/10.1080/10875301.2013.856366.

Havelka, S., & Verbovetskaya, A. (2012). Mobile information literacy: Let's use an app for that! *College & Research Libraries News*, 73(1), 22–23. Retrieved from http://crln.acrl.org/content/73/1/22.short.

Hennig, N. (2014). *Selecting and evaluating the best mobile apps for library services*. Chicago, IL: ALA TechSource.

Johnston, N., & Marsh, S. (2014). Using iBooks and iPad apps to embed information literacy into an EFL foundations course. *New Library World*, 115(1/2), 51–60. http://doi.org/10.1108/NLW-09-2013-0071.

Mackey, T. P., & Jacobson, T. E. (2014). *Metaliteracy: Reinventing information literacy to empower learners*. Chicago, IL: ALA Neal-Shuman.

Mackey, T. P. & Jacobson, T. E. (2011). Reframing information literacy as a metaliteracy. *ACRL, College and Research Libraries*, 72(1), 62–78.

Miller, M. D. (2014). *Minds online: Teaching effectively with technology*. Cambridge, MA: Harvard University Press.

Number of available apps in the Apple App Store from July 2008 to June 2016. (2016, June). Retrieved from http://www.statista.com/statistics/263795/number-of-available-apps-in-the-apple-app-store/.

Perrin, A. (2015, October 8). *Social media usage: 2005–2015*. Retrieved from http://www.pewinternet.org/2015/10/08/social-networking-usage-2005-2015/.

Puentedura, R. R. (2014, December 9). SAMR and the edTech quintet: Designing for learning, designing for assessment. Ruben R. Puentedura's [Blog]. Retrieved from http://hippasus.com/blog/archives/date/2014/12.

Spina, C. (2014). Finding, evaluating, and sharing new technology. *Reference & User Services Quarterly*, 53(3), 217–220. http://doi.org/10.5860/rusq.53n3.217.

Yarmey, K. (2011). Student information literacy in the mobile environment. *Educause Quarterly*, 34(1). Retrieved from http://er.educause.edu/articles/2011/3/student-information-literacy-in-the-mobile-environment.

CHAPTER 6*

# Getting Meta with Marlon
## Integrating Mobile Technology into Information Literacy Instruction

*Regina Lee Roberts and Mattie Taormina*

## Introduction

The last place that you might expect to find the use of mobile devices and asynchronous learning is in a library's archives and special collections department. This is due to the fact that archives, as a place, house and care for rare and fragile materials in both analog and born-digital formats. While computers, mobile devices, and digital collections, all technical developments of the recent past, are included in modern archives, they do not readily come to mind when planning library information and archival literacy workshops in special collections. Yet this is exactly what students find when they enter the Stanford University's Special Collections and University Archives classroom for a combined workshop. From an instructional viewpoint, this chapter highlights several key concepts in information and archival literacy while seamlessly integrating mobile devices as a pedagogical tool. A case study is included to provide an example of how the integration and the use of iPads plays a role in expanding the one-shot workshop into asynchronous sessions with multiple layers of possible learning outcomes. In this particular study, the library instruction session was aligned with discipline-spe-

---

* This work is licensed under a Creative Commons Attribution-NonCommercial-ShareAlike 4.0 License, CC BY-NC-SA (https://creativecommons.org/licenses/by-nc-sa/4.0/).

cific research skills, making the library workshop a meta-analysis of the methods being developed through the course. The curriculum design for this workshop employs methods grounded in, and dedicated to, student-centered and *active learning* pedagogies.* This case study addresses the challenges mentioned in the *NMC Horizon Report 2016 Higher Education Edition* by blending formal and informal learning, improving digital literacy experiences, and cultivating genuine curiosity for students (Johnson et al., 2016).

## Background

The library/archives workshop described here builds upon several years of praxis developing information and archival literacy workshops that included active learning components but heretofore had never fully integrated the use of mobile technology to enhance the student experience. The authors' approach to teaching these workshops is based on the following principles: a library-as-lab model blending information literacy with archival literacy; the use of co-design (Somerville, 2009); and the use of active learning practices to teach research skills (Bahde, Smedberg & Taormina, 2014). These concepts are also core principles of critical library instruction, relational information seeking, and active learning models that are discussed by Accardi, Drabinski, and Kumbier (2010); Kuhlthau (2004); and Bruce (1997). As an important component of curriculum development, the authors strive to create workshops that provide a more inclusive information literacy experience by involving the faculty and graduate student teachers in the planning phases as well. This allows for a better sense of the instructor's goals for the class. This inclusive teaching philosophy is fundamentally influenced by Freire's concept of the learner not as an empty vessel needing to be filled, but as someone with life experiences that inform his or her learning (1970).

This pedagogical grounding requires librarians to create programs that facilitate an expectation that students will be contributors to the class. With this perspective in mind, the librarian's responsibility is to create a special collections workshop that will engage participants in discussions and an active analysis of materials. Thus, in the following case study, students working in small groups highlight and unfold possible research trajectories while modeling research methodologies.

---

\* Active learning is where students engage in some activity that forces them to think, analyze, synthesize, and evaluate information in discussion with other students, through asking questions, or through writing. It means that the students are actively participating, not simply listening to yet another lecture (Stanford University Center for Teaching and Learning, "Active Learning: Getting Students to Work and Think in the Classroom," *Speaking of Teaching* 5, no. 1 [Fall 1993]: 1–3, http://web.stanford.edu/dept/CTL/Newsletter/active_learning.pdf). In these workshops, students are developing skills in applying concepts in their chosen disciplines to archival and rare library materials.

This "model" workshop incorporates many of the metaliteracy or "threshold" concepts mentioned in the 2016 ACRL *Framework for Information Literacy for Higher Education* (ACRL, 2016). Noteworthy frames include the following: authority is constructed and contextual; information creation as a process; information has value; research as inquiry; scholarship as conversation; and searching as strategic exploration. This case study shows that the inclusion of the use of iPads in the workshops enhances the focus on these framing concepts by giving students a tool to actively work with while creating an analysis, locating value of the information, strategically exploring the collections, and questioning the content. Ultimately, the students produce their own conversations and analysis about the collections through a lens that is linked to their specific coursework. Using tools related to mobile technology creates space for an authentic learning practice through reflection. As explained by Kearney and Schuck, "Authentic learning comprises learning in ways that fit with real world contexts, where the learning is motivated and developed by the context, and is also learning that develops skills and concepts for effective living in contemporary society now and in the future" (2004, p. 2).

When selecting items from the collections for use in a typical combined workshop, the authors purposefully look for items that facilitate student response, engagement, and critical thinking related to course-specific content while teaching the underlying principles of information and archival literacy. Items that lend themselves to close reading or observation are especially desirable for hands-on learning experiences. To that end, the selection is usually an array of resource "types" that demonstrates different yet complementary research methodologies. Surveys, correspondence, maps, field notes, ledgers, statistics, and court records are some examples of resources often selected for use. Additionally, artifacts such as a stylus, eyeglasses, a historic computer mouse, and a manual typewriter are included in order to expand on the experience of what one might find in a manuscript collection.

After selecting relevant sources that meet the class's pedagogical goals, unique exercises based on those materials are crafted. The class exercises require students to answer specific questions relating to the selected materials in pairs or groups, and individuals are given time to report their findings to the class in turns. These classes usually conclude with additional time for browsing other relevant items pertinent to their class's topic.

The Stanford University Anthropology Department routinely invites the library to be a part of its research methods courses. Librarians have provided one-shot instructional sessions that have evolved over time to include archival research as a main component. For anthropology methods courses, the associated methods for collecting evidence and supporting arguments for developing and answering anthropological questions is included in the archival exercise questions and discussion topics. A goal is to review the inherent links between how a question is framed, the types of evidence that can address the question, and the ways in which social science data is collected in order to do research. In a typical anthropology methods course, students are required to conduct sample research activities, such as interviewing, participant observation, quantitative observation, archival investigation, ecological survey, linguistic methodology, tracking extended cases, and demography. The library workshop segment of the course features archival investigation utilizing collections that demonstrate anthropological research methods through, among other materials, transcripts of interviews, notes from participant observations, and research notes on language and meaning. In keeping with the co-design principles mentioned earlier, the librarians invite faculty or teaching assistants to brainstorm in advance of the class with regard to source selections and workshop activities.

The case study below synthesizes these pedagogical concepts and information literacy frameworks and the authors' curriculum design process. The workshop described is not just an information/archival literacy workshop. It touches upon, and is grounded in, ethnographic methodologies and research skills that prepare students to do anthropological fieldwork. In this case, it also delves into the discipline-specific research process of making a documentary film by analyzing one particular manuscript collection from the perspective of an ethnographic methods lens.

## Case Study: The Marlon Riggs Collection

In spring 2015, a professor in the anthropology department invited the authors to prepare a library workshop for her methods class. This faculty member was familiar with the authors' student-centered active learning approach to workshops and agreed to meet in advance to discuss the workshop's framing and material selection. In this initial meeting, the faculty member mentioned an interest in getting the students to explore the meanings of performance and performative participation in activism. She wanted to look at archival collections that answered the following fundamental questions of performance activism: What does it mean to create? What are the relationships between thinking and making? She wanted her student to think about how research becomes performance and about visual anthropology methods, such as interviewing techniques and collecting oral histories using video. She also wanted them to think through issues of race, gender, class, and ethnicity.

Immediately, the authors suggested using the Marlon Riggs Papers (1957–1994) exclusively. This one collection had enough rich and diverse content related to the idea of performance activism and aspects of ethnographic documentary filmmaking to be useful for all aspects of the workshop:

This collection documents the life and career of the documentary director, Marlon Troy Riggs, 1957–1994. The majority of the materials in the Collection are from the period between 1984 and Riggs' death in 1994, the decade of his concentrated filmmaking activity, as well as some more personal materials from the late 1970s onwards. The papers include correspondence, manuscripts, subject files, teaching files, project files, research, photographs, audiovisual materials, personal and biographical materials created and compiled by Riggs. (Pappas, 2011, p. 2)

Marlon Riggs was an openly gay African American man. He graduated from Harvard and received a master's degree in journalism from University of California, Berkeley, with an emphasis in documentary filmmaking. He incorporated the use of "participant observation" into his work, a discipline-specific method used in anthropological work. His films continue to challenge racial stereotypes and the systematic silencing of gay black men. Some of the more widely known films that he produced include *Ethnic Notions* (1987), *Black Is... Black Ain't* (1995), and *Tongues Untied* (1989). All of his films address the roots of racism in the United States as well as other biases related to gender and sexuality. Riggs's films are reflexive and revolutionary for their time period. Because he received funding from the National Endowment for the Arts (NEA), his films sparked a national debate about what "types" of projects should receive federal funding. His NEA-funded film *Tongues Untied* celebrated the gay rights movement and highlighted the intersection of race and ethnicity within the movement. Public Broadcasting Service (PBS) stations received such considerable backlash from conservatives across the country that many refused to air the film, and the late Senator Jesse Helms specifically pushed to end public funding of this work.

Due to the richness of the collection, it was decided that the one-shot format would not allow for enough time to build context and follow up in a meaningful way within the regular class time period. Therefore, a flipped classroom approach was incorporated. Students were asked to watch one of the three Riggs films mentioned above in advance of the class so that the time spent in the special collections classroom could be used more deliberately. In class, the librarians wanted to emulate the research process and pose questions such as, what does a person need to do in order to create documentary or ethnographic film, get published, deal with censorship, or create a presentation on a research project? Therefore, workstations with specific themes were organized. The authors wanted students to begin thinking about research as a process, from the gathering of information phase through the creation of a final production.

Ultimately, this led to the incorporation of iPads in the class as a way to further extend the workshop experience. Most students are familiar with making videos either on their phones or with other devices, but this workshop also challenged them to think through the complexity and the full range of steps required to make a documentary film. The Riggs collection provided very real and tangible evidence of this, including grant proposals, background research, producer notes, correspondence, transcripts, and more.

Concepts within the Riggs collection were first disentangled by identifying themes. Then the themes were broken down into parts of a greater whole in order for students to piece them back together through student-led analysis. Six different stations were created in the classroom and focused on the following aspects of documentary filmmaking: ethics, research, production process, business, reception, and a panel discussion.* The class was split into groups of three, and each group rotated from station to station with enough time to view, analyze, and report on each station's unique materials. The students were required to report on at least one question from a list of prompts before rotating to another station. In order to incorporate anthropology fieldwork methods into the activity, the students were required to interview each other and video record those interviews at each station. This allowed students to step into the role of producer/ethnographer interviewing subjects and, conversely, being the subject of an interview. The activity was prefaced with a short discussion about the interviewing process, permissions, and methods of record keeping when recording interviews. In order to do this meta-activity, the student groups needed recording devices.

## Mobile Device Selection

The criteria for selecting a recording device that would meet the discipline-specific learning objectives and technology needs of the class were that it had to be lightweight, small, portable, easy to use, and able to produce easily editable video files for future manipulation. In order to share hat the postproduction files could be shared with faculty and students, the video files needed to be in a format that was uniform and easily transferable to a cloud-based file-sharing system.

The authors decided to utilize mobile technology over traditional camcorders since it satisfied all of the criteria, including the ability to seamlessly interact with a cloud-based file-sharing application called Box (https://www.box.com/home). Initially, the authors considered having the students use their own cell phones to record their archival discoveries, but decided that a more streamlined approach was to have the librarians edit and upload the newly created files to Box after the class so that the students could access the set of films for review later on through their course materials management system. The goal here was to encourage critical thinking about the content after the class and to use technology as a tool to enhance and extend the workshop beyond the classroom. Keeping in mind how

---

* See the appendix 6A for details on the station contents and questions.

busy students are and how much they multitask, it was determined that it would be best if students used library-owned recording devices to record their information instead of their personal phones. This too, facilitated the sharing of complete sets of post-workshop film clips for sharing for each station. Combined video clips were then posted to the course management system via Box, which allowed students to watch and analyze them later, asynchronously. Having the students edit and upload the files themselves was beyond the scope of this workshop, but can be an addition to the workshop in the future.

After framing the workshop's logistics, the authors determined that the use of iPads would achieve all of our technological and pedagogical goals. Since many of our students have personal cell phones, which tend to be iPhones, the authors believed that they could easily make the leap to using an iPad. The familiarity of the devices underscores the point that common tools which many students already possess, could potentially be used for other anthropological field research projects in the future.

The library's academic computing department had two iPads available for library use, and the authors supplemented that number with four more from various library divisions. Since one station was set up to be a panel-type discussion, a tripod was also used to keep the iPad stationary while recording the students' sessions. Each station had a dedicated iPad for the students to use, which kept the clips together by station category. Explanations of the exercise's mechanics were provided to the students at the beginning of the class, but virtually no instructions were given on how the iPad's recording features worked.

The iPad's built-in microphones were sufficient to capture the peer-to-peer interviews, even though they were not optimal in the workshop setting. Although the authors knew that the student recordings would pick up background noise in the workshop classroom, they wanted the students to think about these conditions critically from a methodological standpoint in their future work and to consciously try to control for that. To emphasize this point, students were asked to think about sound quality in general, the lighting, and other technical issues that documentary filmmakers might need to consider. The inclusion of iPads as a technology opened up real contextual examples on why methods matter.

Within each group, members were required to take turns being the "informant(s)" and being the "documentarian(s)/ethnographer(s)." This was part of the metacognitive skills that this exercise promoted. Even though there were six stations, each group had time to visit only three. Thus, it was critical to have the students' video responses from all the stations for the post-workshop review option. The post-workshop review was developed to allow students to see and learn about the stations they had to skip and so that they could analyze their own performances as interviewees and as videographers. This also ensured that everyone's voice was heard.

At each station, students would have time to review the materials and questions as a group. The questions at the stations were meant to be different and open-ended, in the same way that an anthropologist or a documentary filmmaker might conduct an interview. The students were encouraged to come up with their own questions based on what they saw on the tables as well. They were free to decide which roles to take on for the video-recording segment. Each group was given five minutes to record its "interviews" with the iPads. Students were required to go through the permissions process for interviewing human subjects and think about how they would track the names of the people on the clips, especially if this had been their own project. In addition to this, a library protocol that secures permissions to record students and share their work was observed and discussed as a method that students would need to incorporate into their own future interview or documentary work.

What would normally be the in-class follow-up and discussion following an archival class became the edited video clips that were sent to the instructor to be added to the class's course management website. Typically, when the librarians solicit reflections or feedback in class, there is not enough time for everyone to get a chance to report. The iPad videos removed that barrier and encouraged candid student responses about the content found in the Riggs collection. Students were free to watch the combined workstation clips, analyze their methods of interviewing, and hear what their peers had to say on their own and at a time that was convenient for them.

## Reflections

Using mobile technology tools to extend the library workshop enriched the curriculum and made this class session a truly metacognitive and expressive learning experience for the students. The use of iPads for video recording expanded the peer-to-peer analysis of archival materials, methods, and information analysis. Having the clips in a format that was easy for the librarians to merge at the end made it easy to deposit the combined clip files into the course management pages. Pedagogically, this allowed the students a third opportunity to engage with advanced concepts in the workshop.

The use of iPads allowed for a portion of the workshop to focus on media literacy skills as well. The ability to quickly post the videos to a mediated space, like the course materials website, added a layer of networked digital communication. It also required a discussion about privacy rights and permissions to post video interviews as part of anthropology ethics and digital media literacy concepts. This is in alignment with the expressed need to include media literacy in educational settings, as discussed by Meikle (2016). The emphasis on integrating media literacy through the use of iPads in higher education is also supported by the *Proceedings of the First International Conference on the Use of iPads in Higher Education* (Souleles and Pillar, 2014). Surveys and findings at this conference reinforce the notion that iPads enhance active learning, lower barriers for peer-to-peer discussions, and create opportunities for students to express themselves in creative and authentic ways. The incorporation of the iPad combined with cloud-based Box hosting and file-sharing service in the course management pages shifted the workshop and lesson plan experiences to include multidimensional learning opportunities. As noted by van der Ventel and Newman, the iPads allowed class time to be devoted to problem solving instead of lecturing or basic introductions to the subject matter. This was true in this workshop as well because the use of iPads enabled deeper interaction with the content of the collections.

The hands-on, active learning portion of the class was enhanced by the iPads because their use enabled a layer of "expressive technology" that had not been a part of earlier workshop designs. Expressive technology as described by Blikstein (2008) includes the use of technology in order to foster learning and is a key to providing a Freirian learning trajectory (Freire, 1970; Freire & Faundez, 1989). The pedagogical influence of Freire on this curriculum is a foundation for working through information and archival literacy principles. By using technology as a way to interrogate the subject matter, students had an easier path to analyze and assess the materials and to create a new object that reflected their authentic inquiry and voice. This case study is a rich example that can be modified to fit other library and archival workshops in similar ways.

It is the authors' view that the quality of the students' discussions in class and in the clips was elevated and participation also increased. In fact, a course instructor commented that her quietest students were much more engaged in the workshop than usual. The authors attribute this notable engagement to the use of the iPads, which afforded the students autonomy over the discussions. These peer-to-peer interactions lessened barriers to participation. Since the students knew in advance that the clips would be deposited into a restricted course management page, they felt freer in expressing their ideas and conversations on film as well. Maintaining this level of trust was important and adheres to some of the ethics in ethnographic research.

Even though the goal of the workshop was for each student to review what his or her peers said about each station, the recorded videos also allowed the librarians to evaluate and reflect on their own curriculum design and classroom experiences in a way not possible before. This was an unexpected evaluation tool that allowed the librarians to see whether or not they had achieved the benchmarks outlined in the ACRL *Framework for Information Literacy for Higher Education* (ACRL, 2016) in this workshop. While they did not factor in accessibility, disability, and inclusion issues regarding the use of the actual devices, it is recommended that future iterations of this exercise address these issues. Future use of the peer-to-peer interview activity can easily be mapped onto other discipline-specific library/archival workshops.

## Conclusion

This library workshop aligned with the research skills of the specific quarter-long course on anthropology methods, fostering a meta-analysis of discipline-specific skills being developed by the students in the class. The mobile technology used for the workshop session included iPads, cloud-based file management, and an online learning management space for sharing post-production files. The case study demonstrates how to incorporate active learning and flipped classroom approaches. The use of iPads for on-the-fly video capture facilitated a peer-to-peer learning environment and dialogue that promoted metacognitive and information evaluation skills.

The authors are proud to have exposed these students to the richness of the Marlon Riggs Papers, which so aptly epitomize many aspects of applied performance activism found in cultural anthropology:

> Dutifully, nevertheless, I attended classes, in search of something more than knowledge or scholarship—in search of a history, a culture that spoke to my life. A history and culture that, simply, talked to me.
>
> Because of this search I began a lesson that, in truth, I've never stopped learning: when nobody speaks your name, or even knows it, you, knowing it, must be the first to speak it. When the existing history and culture do not acknowledge and address you—do not see or talk to you—you must write a new history, shape a new culture, that will. (Riggs, 1991, p.61)

By deeply engaging with Marlon Riggs's archival sources, students experienced an opportunity to inculcate the spirit of Riggs's work on identity and dignity through active learning and engaged dialogue.

## APPENDIX 6A
# Workshop Stations

The themes and questions for each workshop station were as follows:

## Station 1: Ethics/Anthro

Materials on the table included copyright permissions requested by Riggs for the use of music and images. There were transcripts from interviews by Riggs of bell hooks and others, who are in the films.

- How do these documents disclose community values and connections?
- What are some copyright or rights issues that may need to be addressed and why?
- What evidence is there in the archive that shows that Riggs did this?
- How does the bell hooks information align with Riggs's notions of identity construction?
- What are some of the social issues that Riggs is trying to explore in his documentaries?
- What do you think of the "Self Evaluation"?

## Station 2: Research

Materials on the table included images sent to Riggs from the Library of Congress, secondary research materials on the people he interviewed, personal notebook with research lists and notes, research on history of slavery in the United States, newspaper articles, journal and magazine clippings, etc.

- Based on the archival materials you are looking at, can you sense what was driving Riggs's work?
- What kind of research do you see in this collection, and how might you protect your informants in a documentary film?
- What kind of information do you find in the notebooks? What questions did they raise for you as an anthropologist?
- What is most impressive about the scope of research that Marlon Riggs saved?
- How important and how impactful is it to see historic records that give proof to actions?
- The idea of historic erasure: What happens if there are no archival sources for a group or experience?

## Station 3: Process

Materials on the table included transcripts from video recordings, time lines for sound and video clip merging with handwritten marginalia, cassette tapes of interviews, VHS tapes of video recordings, and recording plans with handwritten notes.

- How did Riggs organize his interview transcripts, and do you have a comment after reading some portion or segments? [Please describe the segment that you read.]
- When reading these transcripts as an anthropologist, what do you think of?
- Why save multiple versions of the transcripts? How do researchers do this today when files are created online with various software? What might a researcher need to do in order to save versions?
- What stands out for you in the "Script Outline" for *Ethnic Notions*, and why?
- In what ways do you literally SEE Riggs's editing process? Please describe what you found.

## Station 4: Business

Materials on the table included multiple versions and years of grant proposals that were successful and also proposal that were rejected, expense reports, information about renting video equipment, recommendation letters, marketing information, press releases, and research on grant opportunities.

- Why might Riggs need letters of support for his films, and based on the evidence, what are some of the organizations that he reached out to? Why might you, as a researcher, look into the history of those organizations or individuals who wrote recommendation letters for Riggs?
- How do the *Tongues Untied* press releases vary, and what can we tell from them?
- What do you think about Riggs's write-up on *Black Is . . . Black Ain't*? How is this write-up related to the business side of making a documentary film?
- Why might credentials matter, and when do you offer detailed credentials and when do you offer abridged versions of credentials?
- What do you notice about the budget? Would a documentary filmmaker need all of these categories today?
- What did you notice about the scope and depth of the grant? How might you find out about granting institutions?

## Station 5: Reception

Materials on the table included letters of support and letters against Riggs's work, news clippings on the NEH debates around *Tongues Untied* and airing on PBS stations, media reviews of Riggs's films, and the brief on the Helms Amendment No. 1175.

- As a researcher, how might you use Riggs's list of media reviews?
- Read and describe one or two of the reviews in the folder.
- What is the brief on the Helms Amendment No.1175, and why did this matter for Marlon Riggs?
- After reading some of the letters, what stood out for you?

## Station 6: Panel Discussion

This was the station that was meant to allow a group to perform a panel discussion based on the videos watched in advance of the class.

- If you were to remake *Black Is . . . Black Ain't* or *Ethnic Notions*, what current content would you include?
- Was the use of singing, dancing, and performance important or distracting?
- In your view, how is Riggs's work related to anthropology?
- How does Riggs play with history and social memory in his films, and in what ways are his films still relevant today?
- In terms of "tradition," "authority," "taboo," or "doctrine," how did Marlon Riggs creatively defy rules or create rupture to social biases?
- What are some of the power dynamics that Riggs's films explore?
- What questions do you have for your fellow panelists on racial bias and racial identity today that could relate to Riggs's works?

# References

Accardi, M. T., Drabinski, E., & Kumbier, A. (2010). *Critical library instruction: Theories and methods*. Duluth, Minn.: Library Juice Press.

ACRL Board. (2016). *Framework for information literacy for higher education*. Chicago, IL: Association of College and Research Libraries. Retrieved from http://www.ala.org/acrl/standards/ilframework.

Bahde, A., Smedberg, H., & Taormina, M. (2014; 2014). *Using primary sources: Hands-on instructional exercises*. Santa Barbara, California: Libraries Unlimited, an imprint of ABC-CLIO, LLC.

Blikstein, P. (2008). Travels in troy with Freire: Technology as an agent for emancipation. In C. A. Torres, & P. Noguera (Eds.), *Social justice education for teachers: Paulo Freire and the possible dream* (pp. 205–244). Rotterdam, Netherlands: Sense. Retrieved from https://tltl.stanford.edu/sites/default/files/files/documents/publications/2008.Book-B.Travels.pdf.

Box.com [cloud based file sharing system]. (2015). Retrieved from https://www.box.com/home/.

Bruce, C. S. (1997). *The seven faces of information literacy*. Adelaide: Auslib Press.

Freire, P. (1970). *Pedagogy of the oppressed* [Pedagogía del oprimido. English]. New York: Herder and Herder.

Freire, P., & Faundez, A. (1989). *Learning to question: A pedagogy of liberation*. New York: Continuum.

Johnson, L., Adams Becker, S., Cummins, M., Estrada, V., Freeman, A., & Hall, C. (2016). *NMC horizon report: 2016 higher education edition*. Austin, Texas: The New Media Consortium. Retrieved from http://cdn.nmc.org/media/2016-nmc-horizon-report-he-EN.pdf.

Kearney, M., & Schuck, S. (2004). Authentic learning through the use of digital video. Paper presented at the *Proceedings of the Australian Computers in Education Conference*, Adelaide. pp. 1–7. Retrieved from http://acce.edu.au/conferences/2004/papers/authentic-learning-through-use-digital-video.

Kuhlthau, C. C. (2004). *Seeking meaning: A process approach to library and information services* (2nd ed.). Westport, Conn: Libraries Unlimited.

Meikle, G. (2016; 2016). *Social media: Communication, sharing and visibility*. New York: Routledge, Taylor & Francis Group.

Pappas, L. (2011). *Marlon Riggs papers, 1957–1994. M1759*. [Finding aid]. Stanford, CA: Stanford University Department of Special Collections and University Archive.

Riggs, M. T. (1957–1994). *Marlon Riggs collection, M1759*. Dept. of Special Collections, Stanford University Libraries, Stanford, Calif. Unpublished manuscript.

Riggs, M. T. (1991). Notes of a signifyin' snap! queen. *Art Journal, 50*(3), 60–64.

Riggs, M. T. and California Newsreel (Directors). (2014; 1987). *Ethnic notions* [Video/DVD]. San Francisco, California, USA: Kanopy Streaming. Retrieved from https://stanford.kanopystreaming.com/node/116237.

Riggs, M. T. and California Newsreel (Directors). (1989). *Tongues untied* [Video/DVD]. San Francisco, CA: California Newsreel. Retrieved from http://www.aspresolver.com/aspresolver.asp?GLTV;2543729.

Riggs, M. T. and et al. (Directors). (2004; 1995, c1995). *Black is...black ain't* [Video/DVD]. San Francisco, CA: California Newsreel.

Somerville, M. M. (2009). *Working together: Collaborative information practices for organizational learning*. Chicago: Association of College and Research Libraries.

Souleles, Nicos and Claire Pillar. 2014. *"Proceedings of the 1st International Conference on the use of iPads in Higher Education (ihe2014)."* Paphos, Cyprus: iPads in Higher Education.

Stanford University Center for Teaching and Learning. (1993). Active learning: Getting students to work and think in the classroom. *Speaking of Teaching, 5*(1), May 05, 2016. Retrieved from http://web.stanford.edu/dept/CTL/Newsletter/active_learning.pdf.

van der Ventel, B., & Newman, R. (2014). "The use of the iPad in a first-year introductory physics course." In *Proceedings of the 1st International Conference on the use of iPads in Higher Education (ihe2014)* (pp. 66–77). Paphos, Cyprus: iPads in Higher Education.

CHAPTER 7

# Clinical Resources for the Digital Physician
## Case Study and Discussion of Teaching Mobile Technology to Undergraduate Medical Students

*Maureen (Molly) Knapp*

## Introduction

Undergraduate medical education represents a special case for information literacy instruction, especially because the undergraduates in this population are generally high-functioning adult learners who possess a bachelor's degree and have completed the first two years of medical school. Students are offered a seminar on mobile technology when they are in their third or fourth year of medical study in the United States and are actively rotating through hospitals and clinics where they observe, learn, and participate in the practice of medicine. Very soon these students will graduate with a medical degree and go on to a residency appointment to "practice real medicine" and hopefully become licensed physicians. What interest should these future physicians have in information literacy? In this case, information literacy instruction is presented as an interdisciplinary seminar on the use, evaluation, and connotations of mobile devices in medical care. The objective of the seminar is to teach students to utilize mobile apps in the practice of medicine with a critical eye towards authority, the value of information, the discovery process, and the use of technology as a substitution for complete medical knowledge mastery.

This chapter will discuss the integration of mobile technology into information literacy instruction in the context of medical education. It describes the development of a monthly one-shot workshop in the Tulane University School of Medicine Interdisciplinary Seminars Series. This seminar has been taught over twenty times in the past three years. Course content has evolved based on student feedback, teacher experience, and the ever-changing app universe. This chapter provides a basic lesson plan for a seminar on mobile resources, reflects upon the effectiveness of various teaching methods utilized in the course, and considers the development of personal app literacy in the context of teaching mobile resources to adult learners.

## *Background*

Clinical Resources for the Digital Physician was created in March 2013 when the library was invited to develop a workshop for the Interdisciplinary Seminars Series (IDS) at the Tulane School of Medicine. The School of Medicine established IDS as a mandatory electives series for medical students in 2008–2009 "to bring students and faculty together to discuss topics that may not be represented, or covered in depth, in the standard curriculum" (Tulane School of Medicine, n.d.). Third- and fourth-year medical students are required to complete a minimum of five IDS seminars prior to graduation. Attendance and evaluation are tracked online through the Office of Medical Education, and workshop feedback is regularly distributed to instructors. The library began offering a monthly, two-hour, hands-on workshop on medical mobile apps in its cozy conference room. Attendance was limited to twelve students per seminar.

## *Why Medical Apps?*

By April 2013, the use of mobile phone apps for medicine and health care was sufficiently robust that subscription database providers were including mobile versions of their resources for individual users as part of library subscription packages. Such resources, known as "point-of-care tools," are common to health sciences libraries and are of particular interest to medical students. Point-of-care tools are designed to provide evidence-based medical information that can be used quickly at the bedside by a health care provider. For example, a point-of-care tool might do things such as provide differential diagnoses for an illness, calculate drug doses, or briefly outline the history, treatment, and outcomes of a disease. Point-of-care tools usually include links to research literature, health-care treatment guidelines, or a list of bibliographic references. Some common point-of-care tools marketed to health science libraries include DynaMed, Epocrates, Micromedex, and UpToDate, but there are many more. The Rudolph Matas Library had a wealth of subscription-based mobile resources and needed a conduit to promote them.

In addition, governmental entities in the United States, such as the National Institutes of Health (NIH), the Centers for Disease Control and Prevention (CDC), and the Agency for Healthcare Research and Quality (AHRQ) were developing apps to support clinical practice guidelines, treatments, and health tracking. The growth of health-care and medical apps from private sector app developers only added more choice to the mix—19,000 apps on health-care topics alone were released in 2013 (Eng & Lee, 2013). The reason for this increase was that the medical app market was, for the most part, unregulated. The FDA did not codify its regulations of mobile apps for health and medicine until September 2013. There was opportunity for conversation, both about what apps were available and about how to evaluate them.

## Course Design

Active learning is a requirement for the interdisciplinary seminars at the Tulane School of Medicine. The workshop designers elected to use the format of group discussions, interspersed with short lectures and app demonstrations, with an emphasis on how to install subscription resources. In the beginning, installing subscription apps tended to make up the bulk of the seminar, as each mobile app tended to have specific and unique validation and installation procedures. As time progressed, installing apps became an informal seminar activity, with students discussing what they liked or disliked about the apps while everyone installed things in the background.

Other instructional methods in the seminar included the use of audience response tools, readings, and discussion. These methods will be discussed below, along with other considerations. From an instructional design perspective, these seminars are intended to be considered short modules that could be tied together in any order or used independently. A description of how the course is currently taught will conclude the chapter.

## Audience Response and Feedback

For the first year of the seminar, audience participation tools were used to gauge awareness of how to do specific tasks on a device and what apps the students were already using. Instead of trying to deal with the classroom clicker method, cloud-based polling software Poll Everywhere was used to collect data. Students responded from their cell phones or computers about topics such as their comfort level with completing various tasks on an iPhone and whether or not they were familiar with various library subscription apps. Every month, the responses were recorded so that students in the next iteration of the seminar could see how peers in other seminars had responded. Apps mentioned during discussion were also recorded and shared as a word cloud (figure 7.1) to further spur discussion.

**FIGURE 7.1**
Word cloud: "What apps are medical students using now?" Based on student self-report from previous seminars, April 28, 2014.

Questions for audience response also gauged basic familiarity with devices, such as:
- I know how to open and close an app.
- I know how to group apps together.
- I am familiar with or have installed the following apps (included list of subscription mobile apps from the library).

As time went on, challenges to the use of technology in the classroom such as spotty wireless connectivity for students' phones and hang time while the polls loaded, prompted a change in format. It eventually became easier to replace the audience response systems with a simple show of hands and discussion. Survey questions and in-class demonstrations of the performance of simple tasks on a phone were also phased out, as course evaluations indicated a reasonably comprehensive basic knowledge among students by 2015. The word cloud was a successful method in sharing student feedback from previous seminars with new groups and providing a crowdsourced, homegrown list of resources mentioned in the seminar.

## Readings and Discussion

Several provocative articles on the role of mobile apps in health care and medical education were published in 2014. Prompted by the explosion of medical apps and the questionable accuracy of many of them, Powell, Landman, and Bates (2014) published an editorial in the *Journal of the American Medical Association (JAMA)* discussing the challenges of locating safe, effective mobile apps in health care. Imagining different paths to establishing an unbiased review and certification process, Powell and colleagues (2014) speculated that "in a few years, the notion of a physician prescribing apps might no longer seem far-fetched" (p. 1852). Indeed, as of 2016, at least one start-up is attempting to market a product that enables physicians to prescribe health apps and devices to patients (iPrescribe Apps, 2016). One idea suggested by Powell and colleagues and carried on by others is the idea of personal "app literacy". Wicks and Chiauzzi describe app literacy as "a bottom up strategy to educate consumers about how to evaluate and interpret their own data in health apps" (2015). The laundry list of qualities to consider in health-related apps was so long, however, that "even a trained clinician might struggle to access all the relevant literature and systematically assess every version of every app in every permutation of user, much less under complex security and privacy issues and synthesize them to make a rational design" (Wicks & Chiauzzi, 2015).

Meanwhile, another conversation was happening within the pages of the *British Medical Journal (BMJ)* on the topic of how technology was disrupting the culture of medical education. In "Put Down Your Smartphone and Pick Up a Book," Tobin admonished the rise of the smartphones, noting that "literacy—the most empowering achievement of our civilization—is being replaced by screen savviness" (2014, p. 1). Presenting this relatively short editorial and asking the students if they agreed provided a jumping-off point for heated and passionate conversations about the use of mobile technology in health-care settings that ranged from the "hidden curriculum" of the devices students experience in the clinic (students perceived as goofing off on Facebook versus looking up a drug or treatment) to whether or not learning styles (the medium by which people learn most effectively) were real, or a merely a bunch of educational pseudotheory. Mueller and Oppenheimer (2014) researched the efficacy of note-taking by hand versus laptops and found that subjects who took notes by hand had better long-term comprehension. Ellaway, Fink, Graves, and Campbell (2014) devised a mixed-methods study on medical learners use of mobile technology at one medical school in Canada, finding that different learners adapt their use of mobile devices to the learning cultures and contexts in which they find themselves. On one end of the medical education spectrum, there are high-ranking physicians bemoaning the rise of the smartphone, while on the other end, there are technology champions studying the effect of the devices on learning. Likely, there are both types of people in the same department at the same university. Students are left to wade through those mixed messages and try to determine the smartest way to use technology for themselves.

## *How the Seminar Is Taught Today*

To begin the seminar, introductions are made around the group. Seminar size ranges from seven to twenty students, depending on the time of year, and the seminar includes students in third- and fourth-year medicine. As everyone may not know each other and it is a two-hour small-group session, students are asked to state their name, their year of medicine and chosen specialty, and, if they are nearing graduation, their residency. Identifying specialties in the group helps to tailor instruction, as does identifying levels of study: an early third-year medical student who does not have much experience in a health-care clinic will have different needs than a fourth-year medical student with one month to graduation. Objectives and agenda for the workshop are shared so everyone knows the expectations.

Figure 7.2 is a lesson plan for the seminar. Notice that the course objectives are mapped to a competency specific to medical education and that they refer to the use of informatics tools and information systems in the clinical setting (Association of American Medical Colleges, 2009).

**FIGURE 7.2**
Lesson plan for Clinical Resources for the Digital Physician.

| | |
|---|---|
| **Course Title** <br> **Department** <br> **University** | Interdisciplinary Seminar Series <br> School of Medicine <br> Tulane University, New Orleans, Louisiana |
| **Instructor Name** | Knapp |
| **Lesson Title** | Clinical Resources for the Digital Physician |
| **# of students** | 9-20 students depending on time of year |
| **Date, Time, Place** | May 2016 |
| **Student Materials** | Mobile device (suggested) <br> Paper copies of readings: Eng & Lee (2013), Powell, Landman & Bates (2014) |
| **Preparation for Class** | Make copies of articles <br> Ensure apps to be demonstrated in seminar are updated and working <br> Ensure teaching device is at full power and updated. <br> Disable alerts for any personalized apps during class. <br> Ensure projector will be in classroom. <br> Remember dongle/AV connector for teaching device. |
| **Course Objectives** | Identify and access mobile apps and resources <br> Discuss current research on mobile devices in healthcare & implications for clinical practice |

**FIGURE 7.2** (continued)

| | | |
|---|---|---|
| **ACRL Standards** | Authority Is Constructed and Contextual<br>Information Creation as a Process<br>Information Has Value<br>Research as Inquiry<br>Scholarship as Conversation<br>Searching as Strategic Exploration | |
| **Discipline-Specific Competencies** | AAMC-HHMI Scientific Foundation for Future Physicians (2009) MS 08: Apply quantitative knowledge and reasoning—including integration of data, modeling, computation, and analysis—and informatics tools to diagnostic and therapeutic clinical decision making. | |
| **Before class begins** | Ensure projector will be in classroom.<br>Review readings and discussion questions. | |
| **Introduction** | Introductions<br>Agenda<br>Course objectives | Time<br>00:10 |
| **Teaching Strategy 1:** | Readings—Eng & Lee (2013), Powell, Landman & Bates (2014). Students read articles in class. | Time<br>00:20 |
| **Teaching Strategy 2:** | Discussion—Either individually or with a partner, select a free app to review from the Eng and Lee article. Or choose another app you use in a clinical setting. Consider the following features: effectiveness, ease of integration into health care systems, reliability and safety, and any other observations. Share comments with the group. | Time<br>00:15 |
| **Teaching Strategy 3:** | Lecture—Regulation & Evaluating Apps | Time<br>00:15 |
| **Teaching Strategy 4:** | App demonstrations and installation (5 library subscription sources, 5 other discipline specific ) | Time<br>00:30 |
| **Teaching Strategy 5:** | Lecture—Health Acquired Infections | 00:15 |
| **Closing** | Wrap up, questions, reflections | Time<br>:05 |
| **Assessment** | Post-class survey distributed via email as proof of attendance by Office of Medical Education | Time<br>00:00 |

## Readings

Two articles are presented for students to read in the seminar. This may be at variance with typical active learning principles and is entirely opposite the flipped classroom model, but it is the best way to ensure that all participants read the material. This is particularly important because the readings introduce to students how mobile apps are affecting the practice of medicine. The first reading is a research article from *Pediatric Diabetes* which surveys the medical apps market for apps related to endocrine disease (Eng & Lee, 2013), the second reading is the Powell, Landman & Bates editorial from *JAMA* described previously (2014). The readings frame the development of the medical app market from both a research perspective and personal opinion. The articles discuss the challenges of mobile apps in medicine, including the effectiveness of apps, federal regulation, ease of integration into health-care systems, and reliability and safety concerns.

Eng and Lee (2013) serves as the place setter for the seminar because it discusses the proliferation of mobile apps in medicine and then goes on to review just a small slice of the health-care market: endocrinology and diabetes apps. A table of apps current at the time of publication (late 2013) is included, which is used in the small-group activity. The article concludes with a discussion of the challenges of mobile apps in clinical medicine. Powell's (2014) editorial, which describes hypothetical methods by which mobile medical apps could be regulated, is used as the second reading because it comes from a widely recognized, authoritative medical journal. Since this is an editorial piece, the conversation about medical app regulation is continued in a series of replies to the editor from other physicians in subsequent issues. This serves as an example of the ACRL frame "Scholarship as Conversation." In addition, the discussion introduces the concept of app literacy.

## Discussion Activity

Following the reading, the students are directed to work individually or with a partner to select a free app to review and to consider the following: the effectiveness of the app in answering a clinical question, the ease of integration of the app into health-care systems, the reliability and safety of the app, and any other observations. Students can choose an app from the research article or select a different clinical app that they are familiar with or have on their device. The important thing is to drive a conversation about how students personally use apps in a clinical setting, as well as to have students share mobile apps that have proven useful to them as medical students, especially in regard to patient care. This type of self-disclosure activity proved quite interesting in one session, when a student described tracking her heart rate over the course of a residency interview on a fitness tracker.

In the ensuing discussion, apps and observations are recorded on a dry-erase board in lieu of using audience response systems. Conversation is generally far-reaching. Students are encouraged to share their opinions, which they willingly do since they are talking about apps with which at least one of them is usually familiar. The point of all of this is to get students to think critically about how to evaluate apps and how to develop their own app literacy. Most people already have rules about what they like and don't like when it comes to apps (for example: they download only free apps, or apps that don't burn their battery, or apps that are easy to use and have a pleasing user interface). The goal of this exercise is to broaden these talking points regarding what makes a good app to include features that are not so obvious, such as privacy, who is providing the information, and clinical accuracy. Overall, the reading and discussion activity is designed to get students thinking about apps, the number of apps that exist, how they are regulated, and how they are used in health care. For a list of recommendations on facilitating discussion, see figure 7.3.

**FIGURE 7.3**
Facilitating a discussion.

| Facilitating a Discussion |
|---|
| • Have a few concrete reading comprehension questions to which participants can speak directly…medical students expect affirmation they have the right answer. |
| • Admit when you don't know…but you can find out. |
| • Be ready to improvise. |
| • Encourage questions but be know when to quit. Allow three to five seconds after asking a question for participants to respond. |
| • Probe broad themes from the readings. For example: How does mobile technology affect you now as a student learning about medicine? How do you envision technology impacting your practice of healthcare 5 years from now, 10 years? Do you agree or disagree with the research and editorials you have just read? Why or why not? |
| • Use questions from seminar encounters to further develop discussion. The first time a student asked "who cares?" after a lecture on devices and infection was a eureka moment, and an incorporated into the lecture. |

## *Lecture—App Literacy*

After the reading and discussion activity, there is a short lecture on mobile app regulation and on current research and products on the market in the United States. The US Food and Drug Administration (FDA) (2015) defines a "mobile medical app" as an app that meets the definition of device in section 201(h) of the

Federal Food, Drug, and Cosmetic Act (FD&C Act) and is intended either to be used as an accessory to a regulated medical device or to be used to transform a mobile platform into a regulated medical device. The FDA regulates only a very small number of medical apps: apps that measure or diagnose something clinical; apps that connect to an existing device type for purposes of control or monitoring; apps that display, transfer, and store patient information; and apps that convert patient-specific medical device data from a connected device. If the app has an "oscope" connected to it (stethoscope, otoscope, ophthalmoscope) or calculates or transmits diagnostic health information, then it must meet FDA approval.

The FDA does not regulate many apps on the market today, including online textbooks, educational tools, patient education, general medical office operations such as coding and billing, generic aids or general-purpose products, Fitbits, and app stores. There are many challenges to app regulation. For example, traditional methods have not adapted to the fast-paced nature of technology, and randomized controlled trials, which are considered the gold standard of clinical research, are expensive and lengthy. In fact, researchers aiming to systematically evaluate apps have found that the applications they were evaluating updated as frequently as every three weeks (Powell, Landman & Bates, 2014). Security is another issue. A 2014 research report from Arxan, a mobile application security company, found that 87 percent of top 100 iOS apps and 97 percent of top 100 paid Android apps had been hacked (Arxan, 2014). Furthermore, of the "top 20 sensitive health apps in the Android and iOS market," four hacked apps had FDA clearance (Misra, 2014). There is also the story of Happtique, a mobile start-up company that attempted to build a business certifying and prescribing mobile medical apps until it was revealed by an independent health IT firm that two certified apps had security issues, including storing sensitive information as text files and not using secure web protocols (Dolan, 2013).

In light of limited regulation, the slow process of research, and a lack of commercial options, what's a person to do? Options exist: reliance on "expert" reviews or websites; searching the published literature (often discipline-specific, and likely quickly outdated); and developing personal "app literacy." With minimal regulation and hundreds of thousands of health apps, there is a clear need for students and clinicians to develop a personal app quality checklist.

App literacy can be considered another facet of metaliteracy. Biomedical literature has done some research on the evaluating of medical app evaluation in the past decade (Prorok, Iserman, Wilczynski & Haynes, 2012; Burdette, Troman, & Cmar, 2012). What these publications lack in timeliness are made up for in an overwhelming number of quality indicators for mobile apps in the health-care setting. Some quality indicators suggested include: multiplatform availability, offline accessibility, inclusion of advertisements, lists of references, policies on locating and rating new evidence, author biographies, links to PubMed, access to clinical images, insurance formulary medications, federated searching, user alerts, date stamps, and cost. However, the author prefers to boil it down to how the evaluation of any source, be it printed book, website, or database, was framed back in library school: identifying the item's scope, purpose, and authority. These three features can be used to frame many quality indicators of app literacy. Figure 7.4 shows the app literacy funnel.

**FIGURE 7.4**
The app literacy funnel.

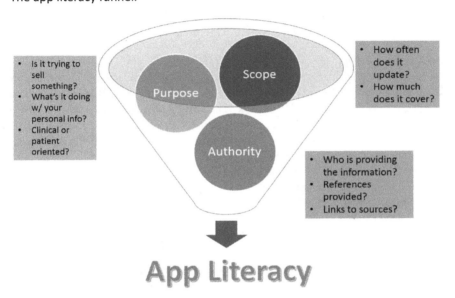

The app literacy funnel is a method to quickly evaluate an app on a personal level and can be easily applied across disciplines.

## *App Demonstrations and Group Installation*

The next part of the workshop is hands-on mobile app installation. For demonstration purposes, an iPad is connected to the classroom projector for this part of the workshop. Five library subscription apps and five CDC health-care apps are discussed, and students are left to decide which to install. Emphasis is on downloading and using subscription apps while they have full access as students. Several of the open-access apps were selected because they fulfill many desirable qualities of clinical apps such as offline accessibility, regular updates, reference lists, and free of cost. A list of apps demonstrated in the seminar is available in appendix 7A.

## *Lecture—Mobile Devices and Health-Care-Acquired Infections*

The final portion of the seminar discusses research about mobile devices and health-care acquired infections (HAIs). Students are asked to consider methods for disinfecting mobile devices, the potential for a mobile device to carry disease, and, ultimately, if this risk is any worse than the risk when using a pen. Research on mobile devices as fomites (physical objects that can carry disease) has been around since the advent of the flip phone (Brady et al., 2011), and more recent studies have investigated how to systematically disinfect mobile devices (Howell et al., 2014; Albrecht et al., 2013). How to address disease on a mobile device is at odds with the Apple product warranty for iOS devices, which recommends they be cleaned only with a lint-free cloth and advises that the Apple product warranty does not cover liquid damage (Cleaning your Apple products, 2016). Howell and colleagues (2014) compared the effectiveness of the lint-free cloth to other disinfectants and found that cleaning agents such as Clorox and antibacterial antiseptic Chlorhexidine digluconate (CHG) were significantly better than the Apple-recommended plain cloth ($P \leq 0.001$) in eradicating common antibiotic-resistant bacteria. In the study, an iPad was infected and disinfected over 400 times, and the functionality and visual appearance of the iPad were not damaged by repeated use of Sani-Cloth CHG 2% wipes (Howell et al, 2014). Manning, Davis, Sparnon, and Ballard (2013) proposed a four-step technique for disinfection: use a waterproof, nonporous case, disinfect the device after a patient encounter, set an alarm to disinfect the device regularly once a day, and use proper hand hygiene as per facility policy for patient interaction. In light of the evidence, some students argued that the danger of picking up germs on a mobile device is no worse than other items such as a pen or necktie. How do smartphones compare? Does it really matter? It was a good question, and one with a history.

In September 2007, the National Health Service (UK) banned neckties in the clinic, citing concerns about health-care-acquired infection. Several studies were published in objection to this new regulation. Pisipati, Bassett, and Pearce (2009) studied the ties and pens of four consultant urological surgeons after five weeks of normal use and found bacteria present in very small numbers. Another small prospective study, by Halton and colleagues (2011), evaluated pens touched by health-care professionals and hospitalized patients with and without cleaning the pen with alcohol-based hand sanitizing agent after each patient visit and found a significant reduction in potential health-care-associated pathogens when the pen was disinfected. Should clinicians be concerned about picking up a nasty supergerm on their phone? The research evidence is currently not conclusive. However, when considering a mobile device as an object that can carry infectious disease, a review of the evidence offers critical thinking points about author bias, study size, and the validity of research results. The populations used in these reviews are small, the authors are possibly unhappy with the new government regulations, and the observed results may not translate to larger populations.

There are three takeaways from the lecture about mobile technology and health-care acquired infections. First, bacteria on the skin and in the environment are not always pathogenic. Any first-year medical student will tell you that we all have the gram positive bacteria *Staphylococcus* living in our nose. Second, the risk of transmission of disease clearly decreases with the use of an alcohol-based sanitizing agent between patients. Even studies that question items as fomites for disease acknowledge that sanitization is effective at eradicating bacteria. Finally, for those who do want to be vigorous in their device sanitization, there is a commonsense protocol: using a waterproof, nonporous case, scheduled disinfections, and proper hand hygiene.

## *Wrap-Up and Reflection*

At the end of the seminar, students have downloaded several new apps and learned methods for thinking critically about app evaluation. As a closing exercise, the students are asked to write down one thing that they have learned or one thing that they still have a question about on a slip of paper. The papers are submitted and read aloud by the instructor. In instructional design circles, this is can be known as a one-minute paper, a reflection exercise, or a summarizing strategy. Regardless of nomenclature, the exercise is an effective method of closing the workshop. Some broad reflections the course instructors have received include "The mHealth universe is daunting," "Take advantage of free library apps now," "Apps complement clinical judgment—they are not a panacea," "Don't believe everything you download," and finally, "Wash your hands."

## *Curriculum Considerations*

From the perspective of library instruction, it is relatively unusual to be able to develop a small hands-on workshop that is given course credit, and it works out nicely because (1) all School of Medicine students must attend at least five of these seminars, and (2) every seminar requires an evaluation to be filled out about the seminar in order to get credit for attending. Thus, student feedback and attendance are automatically calculated and continually provided by the School of Medicine. This feedback contributes to improving and editing the course on a continual basis.

In the context of the classic categories of bibliographic instruction, this course can be categorized both as a one-shot instruction session (for content—students can attend the lecture only once, and they can choose whether to take it from a variety of seminars) and as credit-bearing curriculum-integrated instruction (all medical students have completed a required library module during their first year of study, seminar attendance is credit-bearing). Regardless of classification, the most important learning objective of this workshop is to learn tacit principles of information literacy: the ability to recognize the scope, purpose, and authority of an item and decide its usefulness to the user.

### Lessons Learned

Here are the lessons learned from running a monthly seminar on current mobile apps for over three years. First, teach what you have or what is freely available from credible sources. The primary purpose of bibliographic instruction is outreach to users on library subscription resources. Second, collect and explore recommendations from the user community. Note resources mentioned during the seminar and investigate them. Have a dedicated teaching device preloaded with library subscription resources and other apps. Categorize apps into folders on the device used for teaching. Third, invest in the right dongle for mobile demonstrations on the teaching device, and remember to bring it. Fourth, incorporate and discuss current research evidence and technology news in order to encourage thinking about implications of devices in the user community. In this case, a saved search in the PubMed database for the keyword *mHealth* is delivered monthly to the author's e-mail. Finally, collect, read and respond to course feedback and evaluations. A seminar is not finished once the presentation has been created and saved. This does not mean creating a new class every time, but suggests that for a truly successful and engaging seminar, continual review and adjustment of material and course format on a periodic basis are recommended.

## How Does This Tie into the Information Literacy Framework?

The ACRL *Framework for Information Literacy* (ACRL, 2016) describes information literacy, the understanding of how information is constructed and digested, as a cluster of six interconnected core concepts. This seminar touches on them as follows:

1. Authority Is Constructed and Contextual—When evaluating apps, it is important to ask: Who is providing the information? Is it from a medical textbook? A person? A governmental entity? A copy of handwritten notes from medical school? A drug company? The source may affect the validity of information.

2. Information Creation as a Process—Mobile medical apps often distill medical and drug information from already authoritative sources such as medical textbooks or drug monographs. Emphasizing awareness of the original source of information, as well as how periodically the app is updated and whether it is easy to discover sources and time stamps for updates enhances learning dispositions related to this process.
3. Information Has Value—Some apps cost money; free apps are ingesting user information; all of it is a commodity for someone.
4. Research as Inquiry—Mobile apps for health care are meant to save the time of the clinician, but they are not a replacement for clinical knowledge or patient decision making.
5. Scholarship as Conversation—By delving into replies in editorial articles published in the medical literature, students observe current developments in the regulation of mobile apps, as well as distinguish between editorials and research articles in the medical literature.
6. Searching as Strategic Exploration—There are many apps out there that perform the same tasks. One goal of the seminar is to consider personal preferences for apps and develop an understanding of personal app literacy. There are many paths to the same information.

# Results and Conclusion

As of May 2016, over 200 students have taken Clinical Resources for the Digital Physician. The course consistently receives scores above 4.6 on a 5-point evaluation scale. The course has been offered over three years, and has undergone systematic modification and updating based on student feedback and developments in the rapidly expanding medical app market. As successful a course as it is, this iteration is at a close, as the author is moving on. It is hoped that this chapter serves as a template for others who are passionate about incorporating app literacy into library education and that librarians who are engaged in information literacy instruction will find inspiration here.

## APPENDIX 7A
# List of Subscription and Free Apps Reviewed in Clinical Resources for the Digital Physician App Demonstration

## *Library Subscription Mobile Apps*

AccessMedicine—medical textbooks for mobile devices, wifi not required. From McGraw-Hill Medical. http://mhmedical.com/

Browzine—table of contents monitoring service and journal reader. From Third Iron Advanced Library Technologies. http://thirdiron.com/

Dynamed—point of care, evidence-based information about how to treat disease with copious references, external links to journal articles, and treatment guidelines. From EBSCO Health. http://www.dynamed.com/

Micromedex—two apps are available: Drugs & Drug Interactions. Wifi not required. From Truven Health Analytics. https://www.micromedexsolutions.com

VisualDX—Dermatology and internal medicine differential diagnosis tool. From VisualDx. http://www.visualdx.com/

## *Other Notables*

Centers for Disease Control and Prevention (CDC) mobile apps—a wealth of free, wifi not required apps from a United States governmental agency for both patients and health providers. Resources mentioned in class include: Tickborne Diseases, 2011 Guidelines for Field Triage of Injured Patients, the U.S. Medical Eligibility Criteria for Contraceptive Use ("Contraception"), Vaccine Schedules, and Influenza for Clinicians and Health Care Professionals. http://www.cdc.gov/mobile/mobileapp.html

Epocrates—medical software for drug interaction, EHR, EMR, drug prices, dosing, disease, medical dictionary, ICD9 Codes and Medicare reimbursement. From Athenahealth. http://www.epocrates.com/

ePSS—Electronic Preventive Services Selector—Designed to help primary care clinicians identify clinical preventive services that are appropriate for their patients. Enter in the sex, weight, age of a patient to see screening and preventive services recommendations based on United States clinical practice guidelines. From the United States Agency for Healthcare Research and Quality (AHRQ). http://epss.ahrq.gov/PDA/index.jsp

mobilePDR—the Physicians Drug Reference offers free accounts for health care providers. Access thousands of drug monographs. From the Physicians' Desk Reference® http://www.pdr.net/resources/mobilePDR/

Read by QxMD—table of contents monitoring service and journal reader. From QxMD Medical Apps. http://www.qxmd.com/apps/read-by-qxmd-app

Canopy—medical translator in over twenty languages providing over 4,000 common medical phrases. From Canopy Innovations. http://withcanopy.com/

Figure 1—"Instagram for medical images." Browse and discuss user-submitted medical images. A good example of privacy and security in mobile apps. Calls into question the implications for patient privacy and how apps are collecting and using user-submitted data. These questions are addressed on the Figure 1 website FAQ. https://figure1.com/

# References

Apple. "Cleaning Your Apple Products." Apple Support website. Accessed June 21, 2016. https://support.apple.com/en-us/HT204172.

Association of American Medical Colleges. (2009). *Scientific Foundations for Future Physicians: Report of the AAMC-HHMI Committee*. Washington, DC: Robert J. Alpern, M.D.

Albrecht U. V., von Jan U., Sedlacek L., Groos S., Suerbaum S., Vonberg R. P. (2013). Standardized, App-based disinfection of iPads in a clinical and nonclinical setting: comparative analysis. *Journal of Medical Internet Research, 15*, e176. doi: 10.2196/jmir.2643.

Arxan. (2014). State of Mobile App Security (Research). Retrieved from: https://www.arxan.com/resources/state-of-security-in-the-app-economy/.

Association of College and Research Libraries (ACRL). (2016). Framework for Information Literacy for Higher Education. Retrieved from: http://www.ala.org/acrl/standards/ilframework.

Brady R. R., Hunt A. C., Visvanathan A., Rodrigues M.A., Graham C., Rae C, Kalima P et al. (2011). Mobile phone technology and hospitalized patients: A cross-sectional surveillance study of bacterial colonization, and patient opinions and behaviours. *CMI: Clinical Microbiology and Infection, 17,* 830-5. doi: 10.1111/j.1469-0691.2011.03493.x.

Burdette, S. D., Trotman, R. & Cmar, J. (2012). Mobile Infectious Disease References: From the Bedside to the Beach. *Clinical Infectious Disease, 55,* 114–25. doi: 10.1093/cid/cis261.

Chan S. R. & Misra S. (2014). Certification of mobile apps for health care. *Journal of the American Medical Association, 312,* 1155–6. doi: 10.1001/jama.2014.9002.

Dolan, Brian. (2013, December 13). Happtique suspends mobile health app certification program. *Mobihealthnews*. Retrieved from: http://mobihealthnews.com/28165/happtique-suspends-mobile-health-app-certification-program.

Ellaway R. H., Fink P., Graves L. & Campbell A. (2014). Left to their own devices: Medical learners' use of mobile technologies. *Medical Teacher, 36,* 130–8. doi: 10.3109/0142159X.2013.849800.

Eng D. S. & Lee J. M. (2013). The Promise and Peril of Mobile Health Applications for Diabetes and Endocrinology. *Pediatric Diabetes, 14,* 231–238. doi: 10.1111/pedi.12034.

Halton K., Arora V., Singh V., Ghantoji S.S., Shah D. N., & Garey K. W. (2011). Bacterial colonization on writing pens touched by healthcare professionals and hospitalized patients with and without cleaning the pen with alcohol-based hand sanitizing agent. *CMI: Clinical Microbiology and Infection, 6*, 868–9. doi: 10.1111/j.1469-0691.2011.03494.x.

Howell V., Thoppil A., Mariyaselvam M., Jones R., Young H., Sharma S., Blunt M.,et al. (2014) Disinfecting the iPad: Evaluating effective methods. *Journal of Hospital Infection, 87*, 77–83. doi: 10.1016/j.jhin.2014.01.012.

*iPrescribe Apps.* (n.d.). Retrieved June 21, 2016 from https://iprescribeapps.com/.

Manning M. L., Davis J., Sparnon E., & Ballard R. M. (2013). iPads, droids, and bugs: Infection prevention for mobile handheld devices at the point of care. *American Journal of Infection Control, 41*, 1073-6. doi: 10.1016/j.ajic.2013.03.304.

Misra, Satish. (2014, November 20). Majority of Android and iOS apps have been hacked, including FDA-cleared health apps. iMedicalApps. Retrieved from: http://www.imedicalapps.com/2014/11/majority-top-android-ios-apps-hacked-including-fda-cleared-health-apps/.

Mueller P. A. & Oppenheimer D. M. (2014). The pen is mightier than the keyboard: Advantages of longhand over laptop note taking. *Psychological Science, 25*, 1159–68. doi: 10.1177/0956797614524581.

Pisipati S., Bassett D., & Pearce I. (2009). Do neckties and pens act as vectors of hospital-acquired infections? *BJU International, 103*: 1604–5. doi: 10.1111/j.1464-410X.2009.08440.x.

Powell A.C., Landman A.B., and Bates D.W. (2014). In Search of a Few Good Apps. *Journal of the American Medical Association, 311*, 1851–2. doi: 10.1001/jama.2014.2564.

Prorok J.C., Iserman E.C., Wilczynski N.L., & Haynes R.B.. (2012). The quality, breadth, and timeliness of content updating vary substantially for 10 online medical texts: An analytic survey. *Journal of Clinical Epidemiology, 65*, 1289–95. doi: 10.1016/j.jclinepi.2012.05.003.

Tobin M.J. (2014). Put down your smartphone and pick up a book. *BMJ: the British Medical Journal, 349*, g4521. doi: 10.1136/bmj.g4521.

Tulane School of Medicine. (n.d.) Interdisciplinary Seminar Series. Retrieved June 26, 2016 from https://tulane.edu/som/ome/ids.cfm.

Wicks P. & Chiauzzi E. (2015). 'Trust but verify'—five approaches to ensure safe medical apps. *BMC Medicine 13*, 205. doi: 10.1186/s12916-015-0451-z.

U.S. Food and Drug Administration. (2015, September 22). Mobile Medical Applications [web page]. Retrieved from http://www.fda.gov/MedicalDevices/DigitalHealth/MobileMedicalApplications/default.htm.

CHAPTER 8*

# Mobile Technology Support for Field Research

*Wayne Johnston*

## Preamble

When I arrived in La Paz my prearranged driver failed to show up. After looking everywhere and waiting for longer than seemed necessary, I took a cab from the airport to my hotel. The drive into the Calacoto neighborhood gave me my first sense of the dizzying landscape of La Paz, but even more than that, the street dogs we passed en route struck me. Roaming the streets singly or in packs, picking through garbage for anything edible they could find, every conceivable breed of dog seemed to be homeless in La Paz.

I was in La Paz to meet with researchers at the Universidad Privada Boliviana, who were conducting a series of surveys on the health of children born prematurely throughout the poorer regions of Bolivia. They told me that their previous survey had resulted in five tons of paper, which I could see was taking up almost all of the office space that they had available. The data had to be transcribed from the paper surveys into a database, which required a large investment of time and money. Additionally, errors were introduced during the transcription process. Data validations were then required using the paper surveys, resulting in further delays. The researchers were seeking my advice as they transitioned the data collection to mobile technology in order to achieve greater efficiencies as well as more reliable data.

---

* This work is licensed under a Creative Commons Attribution-ShareAlike 4.0 License, CC BY-SA (https://creativecommons.org/licenses/by-sa/4.0/).

At night I lay in my bed unable to sleep, listening to dogs throughout the neighborhood barking, howling, fighting. It wasn't the noise that kept me awake so much as being drawn into contemplating how a homeless dog problem of this scale could ever be addressed. I learned that there are about 300,000 street dogs in La Paz and the number grows by 20 percent each year. A dog sanctuary big enough to accommodate all the street dogs of La Paz would need to be as vast as the city itself.

A year later, back in Guelph, I met Luz Maria Kisiel, a researcher at the Ontario Veterinary College. Luz was undertaking a research project on the roaming dogs of Villa de Tezontepec, Hidalgo, Mexico. Mexico also has a major problem with street dogs; Mexico City alone reports that it captures and kills 20,000 dogs every month. I worked with Luz to develop a mobile app that enabled her team of local research assistants to locate street dogs, photograph them, record their GPS location, and take notes on their condition. This data supplemented a household survey focused on local residents' experiences with these dogs. Luz's research project will help to inform the first national program of sterilization of dogs in Mexico.

From sleepless nights in La Paz wrestling with a problem that seemed unsolvable to a productive partnership with a researcher back home, I had found compelling examples of the research cycle driven by mobile technology.

# Introduction

In 2009 the author was appointed head of a new team at the University of Guelph Library that was established to support researchers from project conception and grant application through to publishing and long-term preservation of scholarly outputs. It was evident from the outset that there was a range of services and technologies to support researchers within the campus environment, but almost nothing to assist them once they left campus to undertake field research. Virtually every academic department has researchers that rely heavily on field research, whether they are conducting agricultural research in nearby crops or international studies on other continents.

Many researchers rely on paper data collection in the field, which requires that the data be later transcribed into a computer in spreadsheets or databases. This not only delays the availability of the data for analysis but also introduces errors during the transcription process. Paper records are also highly vulnerable to loss and damage. Collecting data using mobile technology, especially when the data is uploaded to a properly administered campus server, makes the data much more secure. The data is immediately available for review and analysis. Among other things, this enables researchers to do data validation while actually on site; often, it is too late to correct errors if the researcher identifies the problems only after she or he is back on campus. The use of mobile technology also introduces additional functionality. This includes more reliable data through the use of authority controls and validity checks. It also enables use of GPS coordinates, photographs, audio, and video in addition to textual and numeric data.

Kevin McCann, Canada Research Chair in Biodiversity and a faculty member in the Department of Integrative Biology at the University of Guelph, provides one powerful illustration of the danger of relying on paper data collection. Like Kisiel, McCann was engaged in an extensive data collection project in Mexico. He and his colleague had to visit many sites, so they purchased an old truck to get them from place to place. All of their data was being collected on paper and stored in the truck. With no scanning of the paper records or transcribing of the data on site, there was no redundancy if anything happened to the original records. They were well into their project and had amassed a considerable amount of data when the truck was stolen. The researchers didn't care about recovering the truck, but they desperately needed to recover the paper, as it represented a huge investment of time and effort. They were very pleased when the police reported that they had found the truck. They were less pleased when the police informed them that they could have the truck back only if they agreed to repurchase it from them. Still, this was a small price to pay in order to get their data back.

It was in this context that the author undertook a multiyear research project to identify ways that an academic library can better support field research. The focus of the project is on data collection using mobile technology, but the research also encompasses data storage, data security, device selection, and other unique challenges encountered by researchers in the field.

## Literature Review and Methodology

An initial literature review revealed that very little attention has been paid to how mobile technologies can be used in academic field research and, in particular, how academic libraries can play a support role. Other aspects of the question were well represented, such as how libraries can deliver services and content via mobile devices or how mobile devices have been exploited for information dissemination in the context of health or humanitarian initiatives. There is some literature available focused on data collection using mobile devices, but generally not from the perspective of academic research.

Gabriel Demombynes and colleagues (2013) present research that features surveys that were administered by a call center. Their report provides an example of using mobile phones for data collection in a developing country where, on the one hand, cell phones are ubiquitous but, on the other hand, network connectivity is unreliable. This dichotomy illustrates why it can be important to employ solutions that capture the data on the device and then automatically upload the submissions when connectivity is eventually established.

In 2015, Michael R. Glass demonstrated how engaging students in data collection using mobile technology can be a valuable component of their introduction to research methodologies. His use of a data collection app for a neighborhood survey provides a compelling example of experiential learning, giving the students a deeper appreciation of field research. Glass cites the particular advantage of having the results of data collection available for immediate analysis.

Sandro Mourão and Karla Okada (2010) describe the development of a mobile solution that features the three main elements that are needed in virtually all tools. These include a mechanism for designing the data collection instrument, a web server where the data will be stored, and a mobile app, which will be used by respondents or survey administrators.

A technology review was also undertaken for this project. This involved identifying software and hardware that could be useful to academic field research. The author read about the tools and resources that had been identified and also acquired them whenever possible in order to gain firsthand experience. When possible, the tools were also introduced to researchers for testing on actual research projects in the field.

This mobile technology research project was informed by a series of conversations with researchers at the University of Guelph. Researchers were initially identified for outreach simply by searching departmental websites. Each conversation included recommendations for other researchers actively involved in field research. Twenty-five researchers were consulted in disciplines as diverse as geography, biology, rural development, plant agriculture, environmental science, political science, anthropology, and population medicine. The conversations were loosely structured to encourage respondents to share stories that would reveal their unique challenges when conducting field research. The topics covered included a description of the environment where the researcher conducts her or his fieldwork, current data collection practices (and any concerns or frustrations associated with those practices), data storage strategies including incidents of data loss, any current use of mobile devices, and what is done with the data upon return to the campus environment.

Although some researchers confessed to being technophobes, the vast majority found the prospect of using mobile devices for data collection to be compelling. Security of the data was a prominent concern. For example, one researcher described collecting data from boats, where the paper invariably became wet and the observations became illegible. Other researchers expressed a need to incorporate GPS coordinates, barcode scanning, or photographs into their data collection work. Some researchers sought tools to assist with sample management, as samples collected in the field need to be associated with identifying metadata. Another common need was the ability to identify data inconsistencies while in the field and to address them immediately, as it is often too late to validate questionable data once the researcher is back home.

In addition to garnering information from the literature and directly from researchers, the author felt it was important to gain firsthand experience by working with researchers in a developing country in order to better appreciate the challenges of working in a remote environment. The author was fortunate to be awarded a research grant from the Canadian Association of Research Libraries that enabled him to travel to Bolivia to work with researchers in three different cities.

The author first worked with researchers in La Paz at the Universidad Privada Boliviana, Centro de Generación de Información Estadística, learning about their experience conducting a survey of about 5,000 households in the poorer regions of Bolivia, the result of which was five tons of paper. They were very eager to migrate their data collection to mobile devices. The second city visited in Bolivia was Cochabamba, where the author engaged with researchers at Fundación PROINPA. While there, the author developed a data collection app to enable an agricultural researcher to measure the growth and yield of her experimental crops. The app addressed concerns such as the collection of data in rainy weather that was problematic with paper, eliminating the need to transcribe data to spreadsheets, and the incorporation of photographs into the data. The third city was Santa Cruz, where the author visited the Centro de Promoción Agropecuaria Campesina (CEPAC). Researchers there were advised on how to migrate their data collection practices from paper to mobile devices.

These experiences in Bolivia tested the conclusions drawn from the earlier conversations with researchers at the University of Guelph and confirmed the great potential for mobile technology to positively impact field research wherever it is undertaken.

## Open Data Kit (ODK)

Many of the applied solutions reported on in this chapter make use of the Open Data Kit. ODK is open-source software developed at the University of Washington with support from Google. It is a suite of tools that enables researchers to create and deploy mobile data collection solutions. ODK has become the core of a number of other products, both commercial and open-source. (Examples include Ona, SurveyCTO, KoBo Toolbox, CommCare HQ, doForms, DataWinners, ViewWorld, and PhiCollect.)

The ODK suite is made up of the following core components:
- ODK Aggregate, which is a server module for deploying data collection instruments, storing the data, and downloading and analyzing data.
- ODK Collect, which provides the data collection app itself on Android devices. ODK Collect stores submissions on the device itself until a connection to the Internet has been established. Enketo extends

deployment to non-Android devices by offering a browser-based alternative.
- ODK Build, which can be used to create the data collection instruments, although XLSForm is a more robust form design tool.
- ODK Briefcase, which is a desktop alternative to Aggregate.

Other ODK modules available include ODK Scan, ODK Survey, ODK Tables, ODK Sensors, and ODK Validate.

XLSForm is a way to create data collection instruments by populating a simple Excel table with the form definitions. While some researchers may not find this approach as user-friendly as a drag-and-drop interface initially, once the principle is understood, it becomes very quick and easy to create forms. XLSForm incorporates controlled vocabularies, value constraints, conditional elements based on previous responses, calculated values, grouping and repeating of elements, and incorporation of media and GPS. Once the form has been designed in XLSForm, it is converted to XForm for use by ODK as well as a range of other platforms.

As mentioned, Enketo is a separate product that deploys the same XLSForm definitions as browser-based alternatives to ODK Collect for non-Android devices. Virtually all the same functionality is retained in the browser version, including the capability to store submissions on the device when not connected to the Internet.

Formhub was developed as an alternative to ODK Aggregate by a group at Columbia University. It facilitated the conversion of Excel form definitions to ODK Collect apps as well as the sharing of forms with other research groups. One of its main advantages was that it offered free hosting and storage for any project that wanted to use it. Unfortunately, the very popularity of Formhub spelled its demise as its use internationally became too much for the dedicated bandwidth as well as the storage capacity. The developers themselves moved on to form a new company called Ona, which resulted in the Formhub code no longer being maintained.

An alternative to installing ODK Aggregate on a local server is to install it on Google's App Engine. There is a well-documented process that makes this easy to accomplish even for users with no expertise in setting up web servers with database back ends. The process for installing an instance on App Engine is essentially (1) make sure you have Java installed on your desktop; (2) set up Gmail and App Engine accounts; (3) download ODK Aggregate and run the wizard-based install; (4) configure the upload of ODK Aggregate to App Engine. You're then ready to begin deploying forms and collecting data with none of the expense and expertise required to install and maintain your own local instance.

Use of Google's App Engine is free for a three-month test period. The pricing after that is difficult to estimate as it involves a combination of transactions and storage. Most modest data collection projects should remain below the threshold where any charges are accrued. Apart from pricing, the other disadvantage of an

App Engine instance is that Google can access the data. For some projects this may not be a concern. However, projects involving sensitive or private data would need an option that is more secure.

## Deployments

There are a number of practical examples of how various tools and strategies have been deployed in actual research projects. As mentioned in the chapter's preamble, Luz Maria Kisiel is a graduate student in population medicine at the Ontario Veterinary College, University of Guelph, and her research interest focuses on roaming dogs in Villa de Tezontepec, Hidalgo, Mexico. Her fieldwork involved recruiting about thirty local research assistants to identify street dogs over a set of five predetermined routes. The author worked with Luz to develop an ODK app that enabled her assistants to take photographs of the dogs, capture their GPS location, and record comments on the dogs' age, gender, and condition. This was supplemented by a household survey of people's experiences with street dogs.

Kisiel was very pleased with the efficacy of ODK Collect as a mechanism for data collection. The project documented and photographed 428 dogs, and she plans to return to the community in a year's time to see if the same dogs can be identified again based on their photographs.

Cameron McCordic is a postdoctoral fellow with the Hungry Cities Initiative at the Balsillie School of International Affairs, University of Waterloo. His research has taken him to Ghana, Kenya, South Africa, China, Mozambique, Jamaica, India, and Mexico for intensive data collection activities. He speaks compellingly of how the use of mobile technology and ODK in particular has transformed their work:

> The use of digital surveys in field research mean that the hardware and software components in tablets and phones can now be integrated into fieldwork more efficiently. These components include the camera, audio recording, and GPS hardware on tablets and phones as well as the added value of downloadable software like random number generators for random sampling, geo-spatial tracking software for enumerator navigation, and encryption software for data security in the field. (C. McCordic, personal communication, April 17, 2016)

Use of this technology has reduced some of McCordic's field research budgets by up to 75 percent thanks to the move from paper to mobile. He also speaks of the efficiencies gained by having data immediately available for analysis and for callbacks to respondents.

One project that presented a particularly interesting set of challenges was a mindfulness research project by Ekaterina Pogrebtsova, a University of Guelph graduate student in psychology working with Professor M. Gloria González-Morales. This project involved extensive use of multimedia, which needed to be available offline and needed to work with all devices. The concept is that respondents will answer a brief survey on their emotional state, go through a mindfulness exercise via audio or video, and then complete another brief survey to determine how their emotional state was affected by the exercise. The researchers want all of this to be available on the respondents' mobile devices so that they can participate wherever they happen to be at the time.

A number of researchers from the University of Guelph have been participating in an international project examining declining fish stocks in Cambodia. One aspect of the project investigated use of mobile technology for data collection. Some challenges were faced with qualitative surveys where the interviews were conducted in Khmer. However, the responses were translated into English during data entry. The researchers were also pleased with the GPS capabilities introduced through the use of mobile devices.

Discovering Biodiversity is a third-year class at the University of Guelph that exposes students to field research. The students are assigned to a plot of land in a forest where they need to identify the trees within the assigned plot. The data had been collected on paper and, given that there are about two thousand students enrolled in the class each semester, this meant a lot of data to be transcribed into spreadsheets before it could be used in subsequent classes. By moving data collection to mobile technology, the data has been made immediately available for analysis.

## Knowledge Base

The primary product arising from all of the aforementioned research is a wiki knowledge base (https://fieldresearch.miraheze.org/wiki/Main_Page). It is hoped that while the author invests time at the outset in populating the site, a broad community of interest will be motivated to further develop the content over time.

The knowledge base is made up of the following sections and subsections.

### *1. Data Collection*

The main focus of the project is on data collection in the field. This section is made up of seven subsections described below.

## 1A. APP CREATION

The App Creation subsection covers tools that facilitate the development and deployment of apps without the need for programming expertise. Included among these tools are those that follow the OpenRosa standard, including ODK.

## 1B. APP DEVELOPMENT

App Development is distinct from App Creation in that it involves designing apps from scratch using frameworks that facilitate the process. While these tools do require knowledge of HTML, CSS, and JavaScript, they generally do not require coding from the ground up. That makes these resources worthy of consideration for research projects with highly specialized needs.

## 1C. SURVEY SOFTWARE

Survey Software covers some of the leading providers of web survey solutions, both open-source and commercial. These are useful for research projects that want to move away from paper and can rely on Internet access without the need for any functionality specific to mobile devices, such as GPS or photographs as data elements.

## 1D. SMS

A selection of SMS tools offer an alternative approach to data collection based on text messaging. Simple, low-cost data collection is advantageous especially when input from a large and diverse population is needed.

## 1E. MOBILE DBMS

A subsection on Mobile DBMS covers database applications stored on the mobile device.

## 1F. SPECIALIST DEVICES

Specialist Devices provides a sample of the wide range of devices that have been custom-built for very specific purposes. These include devices that monitor environments such as air, water, and soil.

## 1G. SPECIALIST SOFTWARE

Similarly, the Specialist Software subsection is meant to give researchers a sense of the range of software options available to address very particular data collection needs.

### 2. Mobile Mapping

While capturing GPS coordinates is a feature of many solutions, Mobile Mapping covers tools that facilitate the capture of more complex mapping data.

### 3. Data Storage

Data Storage is a section that deals with safe and secure storage of data when away from the campus environment. This includes both cloud storage solutions as well as tools to enable data to be sent back to the campus servers.

### 4. Data Transfer

Data Transfer provides information and best practice for the transfer of data between the source and the chosen storage system.

### 5. Data Security

The Data Security section covers additional recommendations on ensuring the security of data, especially when private or sensitive human data is involved.

### 6. Device Protection

Researchers in the field also need to be concerned about protecting the devices themselves. This includes protection from damage when working in hostile environments as well as protection from theft and hacking.

### 7. Data Access

The final section deals with use of mobile devices to access data from the home environment. There are times when researchers in the field will need to consult data collected in the lab or on previous field trips.

## Conclusions

Unlike many technology services, support for field research cannot be provided with any single enterprise solution. This is partly because the needs of research projects across the disciplines are so diverse that the specific requirements of each project need to be matched with a tool that will respond to those needs. This process can include issues of functionality specific to mobile devices, issues of scope, and issues of data sensitivity. It can vary depending on what specific devices are available for use and on issues of complexity and technical sophistication. Another issue to consider with respect to commercial solutions is that many larger data collection initiatives depend on local recruits, which would violate the terms of most commercial license options.

Instead of looking for a one-size-fits-all solution, librarians providing support to field research need to be aware of the range of options so as to make appropriate recommendations. This, of course, is not a one-time acquisition of knowledge. It is critical to keep up-to-date on developments that impact the options available to researchers.

The investment of time required to gain and maintain this knowledge is worthwhile, as libraries that assist researchers with their data collection needs will have early buy-in for broader research data management objectives right through to the publication and preservation of research data. For the most part, no particular technical expertise is required to implement effective mobile data collection solutions. The main challenge is the investment of time to become familiar with the tools and resources, which in turn is diminished when libraries work collaboratively to maintain a shared knowledge base.

While collaboration between libraries is critical, so too is collaboration with campus partners in our own institutions. Campus IT, the office of research including research ethics, and departmental research managers all contribute pieces of the puzzle. Sometimes the library's most valued role is the familiar one of liaison, not just between the faculty member and library services but also with other campus partners. Researchers are often unaware of the support services available, or they are daunted by the challenge of knowing who to contact. Whether the questions relate to data storage, backups, encryption, or ethical vulnerability, the library can ensure the researcher has access to all of the necessary support he or she requires.

# References

Demombynes, G., Gubbins, P., & Romeo, A. (2013). *Challenges and Opportunities of Mobile Phone-Based Data Collection : Evidence from South Sudan* (Policy Research Working Paper No. WPS6321). The World Bank, Africa Region, Poverty Reduction and Economic Management Unit. Retrieved from http://www-wds.worldbank.org/external/default/WDSContentServer/IW3P/IB/2013/01/17/000158349_20130117102554/Rendered/PDF/wps6321.pdf.

Glass, M. R. (2015). Enhancing field research methods with mobile survey technology. *Journal of Geography in Higher Education*, 39(2), 288–298. http://dx.doi.org/10.1080/03098265.2015.1010144.

Mourao, S., & Okada, K. (2010). Mobile Phone as a Tool for Data Collection in Field Research. *World Academy of Science, Engineering and Technology*, 70(43), 222–226.

CHAPTER 9*

# From Start to Finish
## Mobile Tools to Assist Librarian Researchers

*Mê-Linh Lê*

## Introduction

Mobile apps are a pervasive part of daily life. Well over 100 billion apps have been downloaded since 2009, and there are over one million apps that are available to anyone with a mobile device. While librarians frequently use mobile devices and apps for their professional practice, reference, and teaching, the growing ubiquity and utility of these tools strongly suggest their potential for use in another area of academic life: conducting original research. This aligns perfectly with the growing recognition of librarians as researchers themselves, and the associated requirement by more and more academic institutions that librarians perform research as part of their job description. This chapter focuses on the stages of the research process, including project conception, preparation, start-up, data collection and sampling, data analysis, dissemination, and data storage. The best apps (or mobile websites) available to assist librarians as they complete these steps will be reviewed. Factors considered will be usability, design, features, cost, and privacy concerns. This chapter goes beyond commonly recommended productivity tools such as Dropbox or Evernote and will highlight apps that can assist librarian-researchers in the completion of an entire research project from start to finish. A variety of apps are explored, including those designed specifically for the educational market, those created for broad social science research, and those intended for commercial market use.

---

* This work is licensed under a Creative Commons Attribution-NonCommercial 4.0 License, CC BY-NC (https://creativecommons.org/licenses/by-nc/4.0/).

## A History of Apps

Mobile applications, or apps, have been around almost as long as the first mobile devices. Beginning with PDAs and progressing through iPods, apps were initially preloaded on a device and limited to games like Snake or Solitaire or basic tools such as calculators or calendars. The launch of the iPhone in 2007, and the announcement that outside developers could develop apps for it, changed everything. When the Apple App Store launched in July 2008, it had 552 apps, of which 135 were free (Strain, 2015 (Feb 13)). Within a single week, ten million apps had been downloaded (Cohen, 2008 (July 14)); within sixty days, 100 million apps had been downloaded, and over 3,000 apps were available (Cohen, 2008 (Sept 9)). Between 2008 and 2010, Google Android Market, BlackBerry World, and Windows Phone Store all launched—opening up new markets for app developers (Strain, 2015 (Feb 13)). In 2010, *app* was voted word of the year by the American Dialect Society (American Dialect Society, 2010). By 2013, there were over one million apps in the App Store, with sales topping $10 billion USD (Apple, 2013; Ingraham, 2013 (Oct. 22)). By its sixth anniversary in 2014, the App Store hit 75 billion downloads (Dilger, 2014 (July 10)). In 2015, it was estimated that Google Play (formerly Android Market) had twice as many downloads as the App Store—over 200 million during the year (Woods, 2016 (Jan 20)).

All of the preceding illustrates the fact that apps have become a part of our day-to-day life and a primary way that most users interact with content on their mobile devices. The types of apps vary widely. Games are by far the most popular category of apps downloaded from the App Store (22.99% of all downloads) (Statista, 2016). Apps that are potentially more relevant to librarians include those found under Education (9.26%), Productivity (2.77%), and Reference (2.24%) (Statista, 2016). There are now apps available that can compete with many of the traditional tools that librarians might use to complete their research.

## The Librarian as Researcher

The concept of librarians as researchers is one that has been growing steadily over the last decade. The American Library Association (ALA) and the Canadian Association of Research Libraries (CARL) both identify research as a core competency for librarians (American Library Association, 2009; Canadian Association of Research Libraries, 2010). Within the literature, the topic of librarians as researchers, as well as the barriers, challenges, and research supports offered to librarians, have been explored in depth (Berg, Jacobs, & Cornwall, 2013; Hall & McBain, 2014; Kennedy & Brancolini, 2011; Koufogiannakis & Crumley, 2006; Powell, Baker, & Mika, 2002; Watson-Boone, 2000).

Within Canada, the development of the librarian-as-researcher culture has been aided in no small part by the creation of several organizations and institutes aimed at supporting these librarians. In 2012, CARL launched its Librarians' Research Institute to serve as a place for librarians to develop their research skills, work on their research in an intensive setting, be given mentorship, and engage in networking opportunities (Canadian Association of Research Libraries, 2016). As a result of its success, similar programs were started at both McGill University and Concordia University (Carson, Colosimo, Lake, & McMillan, 2014). In addition, the Centre for Evidence Based Library and Information Practice at the University of Saskatchewan was launched in 2013 as a resource to support librarians as researchers. While it was focused primarily on supporting U of S librarians, its annual symposium, resource pages, and blog are extremely helpful to all librarians working on research (C-EBLIP, 2016).

The level of discussion in the literature, the inclusion of research as a core competency, and the creation of institutes and workshops all clearly demonstrate that research is a vital and essential part of an academic librarian's professional practice and is required for career advancement via promotion or tenure.

## *Mobile Tools and the Researcher*

Whether as part of their research or simply as part of their daily professional practice, librarians must constantly seek to stay on top of trends when it comes to information technology, scholarly communication practices, and ways of sharing information. This is particularly true in academic libraries, which have a constant turnover of new students and a strong focus on research programs and scholarly output. As a result, academic librarians must constantly adapt to the changing technological preferences and information-seeking behavior of our patrons.

As has been shown, our patrons increasingly exist and interact in a mobile environment. In order to stay current, we must follow suit. While work on the use of apps in research is limited, and discussed below, there has been some work on how researchers use Web 2.0 tools (nearly all of which are available in mobile form). In a work published in 2009, Kalb, Bukvova, and Schoop explored the research process and the role that social software, such as wikis, microblogging, and social bookmarking, play. They outlined five activities in the individual research process that can benefit from social software: exploration, retrieval, reading, writing, and dissemination, and linked specific tools that can enhance or support these activities. Other studies have explored how academics use Web 2.0 tools in their scholarly work, including research. The majority of the findings indicate that while Web 2.0 tools are being used by researchers in limited numbers, they are not necessarily being used during the complete research process. Instead, these tools are being used in only very specific areas, such as managing citations (e.g., Mendeley, CiteULike, and Zotero) and scholarly communication (e.g., Twitter) (Al-Aufi &

Fulton, 2015; Calvi & Cassella, 2013; Haustein, Bowman, Holmberg, Peters, & Larivière, 2014; Procter, Williams, & Stewart, 2010).

In general, librarians seem much more familiar with Web 2.0 tools than their professorial counterparts, in part due to the ease with which these tools can be incorporated into library programming and reference. In regard to mobile devices and apps, librarians do use them for their professional practice and teaching (Aiyegbayo, 2015; Duncan, Kumaran, Lê, & Murphy, 2013; Smith, Jacobs, & Lippincott, 2010). However, the growing ubiquity and utility of these tools strongly suggest their potential for use in another area of scholarly work, namely research. This is an area where librarians could take the lead, as it is clear very little work has been done on the use of apps to aid in the research process.

Most research that discusses apps focuses on the evaluation of an app for a specific purpose, such as apps for smoking cessation, apps to help consumers monitor tick bites, or apps to measure vertical jump performance (Bricker et al., 2014). While various lists of "must-have" apps abound, they tend to usually be quite wide-ranging in nature (including apps for both personal and professional life). Hennig and Hennig and Nicholas have done excellent work at highlighting apps for use by academics and librarians (Hennig & Nicholas, 2014; Hennig, 2014a). Both works have taken a broader view and included whole chapters on apps for productivity, reading, and note-taking. While both titles do include chapters on research, they do not focus on the research process per se (e.g., data collection, sampling). It is hoped that this chapter will further expand our knowledge of the use of apps by librarian researchers and that its findings will be applicable to all researchers.

## *Why You Should Use Apps for Research*

It is likely inevitable that one may ask oneself *why* a librarian would want to use mobile tools to conduct research. Why would you want to use a small screen with limited typing capability to complete massive research projects when you could just use a laptop? There are several key reasons worth highlighting:

1. Always Available—The amount of research being published these days is increasing exponentially. Librarians conducting research or on the tenure track will be all too familiar with the publish-or-perish pressure. In some cases, it may become the defining feature of one's workday. As such is the case, why not attempt to conduct research using the device that goes with us everywhere at all times? If you think of an interesting idea late at night, you can simply record an audio note. Perhaps you come across an interesting book in the bookstore and want to see whether the library has it. Now you can pull up the mobile-friendly catalog to check.
2. One Device—Depending on the type of research being done, researchers may need survey instrumentation, cameras, notebooks, recording

devices, transcription services, word processing software, and so on. Instead of having multiple devices of which each performs a single function, the use of a mobile device and the use of apps means that one single device can be used to accomplish all these tasks.
3. Apps ≠ Phones—While the inclination may be to link the use of apps with the use of a smartphone, this is not necessarily the case. Apps can be used on desktops, but more importantly they are also available on tablets, such as the iPad and Surface. Tablets now come with keyboards and styluses and are able to give users the convenience of a mobile device with the power and ease of a laptop or desktop computer.

## *The Research Process*

While the research process as written about in handbooks and methodological frameworks is often portrayed as a constant iterative loop, for the purposes of this chapter more practicality is required in order to see a researcher move through all (or most) of the stages of research to successfully complete a project. As a result, this chapter will deal less with theory-based design and more on a progression of steps that includes the very necessary (but less discussed in handbooks) stages of research, including writing, creation of figures, and dissemination.

## *Evaluation of Mobile Apps*

With over a million apps available to consumers, it can be difficult to determine which apps are actually any good. Within the field of education, several rubrics have been developed to evaluate the quality of an app, with a specific focus on their use in the classroom (Walker, 2010). Hennig (2014b) provides a checklist for librarians writing app reviews. This is an extremely comprehensive list that is a bit too in-depth for the purposes of this chapter. Instead, key factors from the checklist are considered, including

- Currency—Updates from 2015 or 2016 were preferred; many apps that were initially considered stopped updating in 2012 or 2013.
- Cost—While free, freemium, or low-cost apps were preferred, there are some instances where more expensive apps were the only or best option available.
- Platform—Apps or tools available on iOS and Android were preferred, but this was not always possible.
- Syncing—Ability to sync across devices is nearly a standard requirement.
- Accessibility—Apps must have been available for download by both Canadian and US researchers.

- Importing and Exporting Options—The ability to pull in data or information from other sources and then transfer that information elsewhere as needed is extremely important.
- Design—Apps that were visually appealing and intuitive to use were preferred.

It is also worth noting that apps are discussed in the context of the activity that is seen as their primary purpose. In other words, while an app such as Dropbox may be mentioned as a tool for archiving, it would nonetheless also work for collaboration and writing. Finally, in almost all cases there is no pricing information or website information listed. While the author was conducting research for this chapter, it became clear that app prices, features, and even their availability can change very rapidly. Rather than including detailed charts or tables with this information that would become outdated by the time this chapter is published, librarians are encouraged to use their very best Googling skills to find the most up-to-date information on the apps they are interested in pursuing.

# Mobile Apps for Research
## *Project Conception and Design*
### MIND MAPPING

While some researchers prefer to begin research projects by writing all their ideas down, others will want to create mind maps as a way to visually organize their information. There are many expensive, although very well-designed, apps available for power users. For those looking for cheaper options, however, the author found the best app to be Coggle, a free web-based application that allows for collaborative mind mapping using one's own design or uploaded images. Maps are easily exportable and can be embedded into other locations very easily. SimpleMind has a free app that allows you to create maps on the go. Syncing to other devices has a cost, and functionality is often somewhat limited. The Total Recall app allows you to create up to three free maps of unlimited sizes and with an array of colors and images to use. For those willing to pay, MindNode and Popplet are two of the best-designed tools in the lower price range.

### NOTE-TAKING

Note-taking is one of the areas in which app users are spoiled for choice. The perennial favorite, Evernote, allows for note-taking to be taken to the extreme. (Users can upload notes via audio, scanning via its Scannable tool, by taking a picture of handwritten notes on its specialized Moleskin notebooks, or even by plain old typing). Evernote notes can be accompanied by checklists and photos and can sync across devices. OneNote also seems very popular, even for those who do not regularly use Microsoft Office. It allows one to take notes, make to-do lists, and create project folders, and it comes with a large range of formatting options. Audio, sketches, and video can be included, and it can be synced across devices. The Google Keep app, which has the appearance of a sticky note, allows one to create notes with checklists and images, write up longer documents, and also to sync across devices. The app has an audio feature that will transcribe any voice note that one records and make it into a note right away. It also fully interacts with the entire suite of Google Drive products. For those who like an extremely simple interface, Simplenote offers a straightforward design and the ability to collaborate on notes and to search through one's whole note history to find something that may have been deleted. For researchers who prefer to dictate their shorter research notes or thoughts, Dragon Dictation allows one to create short audio messages that are immediately transcribed to text and saved.

## BACKGROUND RESEARCH

As librarians know very well, any type of research will require a visit (either in person or electronically) to the library. Most libraries now have mobile-friendly or specifically designed apps to search their OPACs. The list of vendor-supplied apps that search only within their products is extensive. Some notable examples include OvidToday, which allows for full-text access (via your library subscriptions) to all Ovid journals via its app. JSTOR's app lets you search its holdings, access full text, and pull out citations. Within the humanities and social sciences, EBSCOhost's and ProQuest's apps both allow one to search through their databases and access the full text. If your library's institutional holdings include them, EBSCOhost includes access to Library, Information Science and Technology Abstract (LISTA) and Library and Information Science Source (LISS), while ProQuest provides access to Library and Information Science Abstracts (LISA). Combined, these three databases make up the major sources of library science research.

## KEEPING CURRENT

Keeping current with the literature is an area of importance for all researchers. BrowZine is free to the user and available at many institutions. The users simply select the journals that are of interest to them, link to their institutions' own holdings for full text, and then are notified whenever a new issue is released. Researchers can then annotate articles of interest to them and export them, either as a PDF or to a citation management program. For those librarians conducting research in the health sciences, Read by QxMD, Docphin, and DocWise are also available.

## DATA MANAGEMENT PLANS

The completion of a data management plan is becoming a best practice for those completing research for which large amounts of data will be created. Unfortunately, data management is an area for which there is not yet strong support via either app or mobile-friendly sites. Instead, sites that help researchers create data management plans, such as DMP Assistant (Canada), DMPonline (UK), or DMPTool (US), have to be accessed via regular websites and are not necessarily optimized for smaller devices. On larger tablets, however, the tools are often straightforward to use.

## *Project Start-Up*

### MEETINGS WITH COLLABORATORS

If you are working as part of a team, or just need to meet with an interviewee in another location, there are free apps available for meeting online—although not all of them have the full capability of their browser-based product counterparts. join.me is one of the better free apps for online meetings as it allows for video, audio, whiteboard, annotations, and document and screen sharing from within the app with up to ten people. VSee lets you talk with up to five people while sharing your screen using very secure technology, while Google Hangouts enables you to video chat with up to ten people. Both VSee and Hangouts let you share documents and your screen only when the meeting organizer is using a desktop version. Unfortunately, one of the other well-designed products, TeamViewer, has the same problem. TeamViewer is a robust tool and has a free app for joining meetings. The initial meeting, however, must be created using the desktop or web version. In addition, the organizer must be on a computer in order to share documents.

### PROJECT MANAGEMENT

Researchers, whether working on a small team or alone, may find it difficult to keep track of all of the different tasks that must be done in order to complete a given project. There are several tools that provide assistance in the setting of goals, time lines, tasks, and scheduling. Asana, a web and mobile app, allows you to track all aspects of your work, who is involved and responsible for what, time lines, and when items are accomplished. Team members can have conversations within the app that are saved for future reference. The basic version of Asana is free for teams with fewer than fifteen people. Trello is a project management app that uses a card- and boards-based system to manage everything from complex projects to shopping lists, whether on your own or part of a larger research team. Freedcamp is a completely free project management tool that aims to help schools, educators, and small businesses manage their projects. Multiple people can work together;

add milestones, tasks, and scheduling; and have discussions. While it is currently only a web-based program, an iOS app is in the works.

## EXPENSES AND BUDGETING

For those librarians who receive research funding, staying on top of expenses can be a task that gets left until the end of the project. However, keeping track of things such as mileage, travel expenses, and software purchasing costs can easily be kept on top of by using apps designed to keep track of business expenses. Smart Receipts is an open-source app that lets you scan receipts with your phone, and it can handle multiple currencies and generate reports that can be e-mailed or exported. The Shoeboxed app allows you to organize receipts, which can be scanned using your phone (the app allows for text recognition), uses your phone's GPS for mileage tracking, archives receipts automatically, and sends expense reports using your scanned receipt images. The free version of this app allows you to include up to five receipts per month. For librarians whose institutions use Concur as their travel management software, the free app Concur for Mobile can be used to track business expenses and receipts manually or, using your phone, to link to business credit card expenses and to submit and approve expense reports. For librarians looking for a more robust budgeting tool, Mint works well in both Canada and the United States, it links with most major financial institutions, and it compiles detailed budgeting information based on your spending habits. The Goodbudget app allows you to allocate expenses to specific "envelopes" to track spending and share the budget with colleagues.

## *Data Collection and Sampling*

### RANDOMIZATION

There are several tools available for randomization—whether it is of study participants prior to starting research or simply choosing a random prize winner from among survey participants. Research Randomizer is a free web-based program that is designed for researchers and students to be able to generate random sets of numbers. For librarians who may be part of research teams involving patients, Randomizer for Clinical Trial Lite is a free patient randomizer app that allows you to input patient information and follow randomized patients throughout the study period.

### DATA COLLECTION

#### *Surveys*
Surveys are another area of strength when it comes to app availability, although finding one that meets all your needs for free may be a challenge. SurveyMonkey,

one of the most widely used survey tools in library research, has an app that allows you to build surveys, collect responses, and analyze results. The free version is limited to ten questions and 100 responses. SurveyLegend is a free tool that works best in the app environment. You can easily create surveys on your device using a range of well-designed templates and a wide variety of question types. Surveys' responses can be analyzed and exported from the app as well. The first three surveys are free, and you get unlimited responses. Unfortunately, Canadian librarians concerned about where their data is housed are no longer able to access FluidSurveys at a reasonable price. The company has been bought by SurveyMonkey, and storing data in Canada now costs $119 USD per month. For researchers comfortable with more basic surveys, Google Forms is an option via the web-based tool. Users get unlimited free surveys and have access to a variety of themes and question types. Results can be easily analyzed and exported.

## *Interviews and Focus Groups*

For librarians conducting interviews or focus groups, there are various tools available depending on what type of information needs to be captured. For written notes taken during an interview, refer to the category Writing below. For in-person audio-only interviews, Pio Smart Recorder is a free tool that lets you mark during a recording when something particularly important was noted. Voice Record Pro has an easy-to-use interface and allows for basic editing within the app. AudioNote Lite lets you make audio recordings and then attaches written notes to specific points with the recording—a great feature for noting particular areas of importance. QuickVoice Recorder is a free voice recorder app that allows you to switch between the app and other applications while recording. This is especially useful if you want to make notes or look up additional questions while an interview is being recorded. If the interview is being done over the phone, a fairly inexpensive option is TapeACall Pro, which allows you to record incoming and outgoing calls. Calls are secure and can be shared among colleagues. If the interview is to be captured using video, almost all smartphones and tablets have a built-in camera that allows for the capturing of video. The length of the recording will vary based on the phone and the storage capabilities, but recordings of thirty to sixty minutes should not be a problem. Researchers looking for a higher-quality video recording app with more functionality may have to be willing to spend a bit more money. The Camera Plus Pro app captures video and images, allows for editing within the app, and lets you choose the image resolution.

While there are a large number of resources that help researchers run focus groups or conduct market research, many of them, such as Upinion or Liveminds, are prohibitively expensive for most researchers. However, FocusGroupIt is a web-based tool that is simple to use, and the free account offers you one active focus group with up to ten participants.

## Transcription

The transcribing of interview recordings is a time-consuming and onerous task. While no app is yet available that offers free and reliable long-form transcription, there are several that offer starting points. Dragon Dictation is one of the most popular transcription apps but focuses mostly on transcribing audio to text for things like e-mail, texting, or updating a status. While fairly accurate and free, it deals only with live recordings. TranscribeMe allows you to record within the app or import from other sources and offers a claimed 98 to100 percent accuracy rate for $1.49 USD per minute. Apps like Dictamus and Philips Dictation allow you to make secure recordings but require you to have access to a more robust transcription program on your computer.

## *Data Analysis*

Data analysis is one area to which app developers need to devote more attention. Likely due to the fact that data analysis software programs, such as NVivo or SPSS, are often very expensive and complex tools, there are no great low-price apps or mobile-friendly sites available yet. However, since the bulk of librarian research does not produce massive amounts of data (in comparison to, say, medicine or psychology), less robust tools may be acceptable. One such tool is Dedoose, which is a web-based program begun by academics that is specifically designed for mixed-methods and qualitative data analysis. Pricing is based on a monthly model, and it is designed to be particularly intuitive to use. It allows for multiple research team members to work on the same set of information (although individual access privileges can be set for each team member) and is particularly strong when it comes to collaborative features. This app could be particularly helpful for multi-institution librarian research teams. Researchers performing statistical analysis can use StatsMate—an iOS-only statistics calculator that can generate descriptive statistics; conduct hypothesis tests; and provide correlations, simple regression, and one-factor and randomized block ANOVA models (Lomax, 2013). In a study on statistical accuracy, when compared to other apps such as Data Explorer, Statistics Visualizer, and TC-Stats, StatsMate was found to be the most statistically accurate and comparable to Excel 97 (Lomax, 2013).

## *Writing and Citing*

### SCANNING

Whether researching at an off-site location or working out of a foreign library, researchers may encounter the need to scan documents. While previously one would need access to a stand-alone scanner or photocopier for this purpose, there are now apps designed to take a picture of a document, image correct it, and convert the image to a PDF. Genius Scan is a free mobile app that easily scans documents or receipts and converts to a specific format, which can then be e-mailed, archived, or exported (this comes at a cost). The Google Drive app for Android also allows for scanning and direct uploading to Drive and has basic text recognition. For people already using Evernote, the add-on app Scannable quickly scans and edits documents, business cards (which it can translate into LinkedIn contacts), receipts, and more. Other notable free scanning apps include CamScanner and CamBot.

## WRITING

One activity for which there are a multitude of high-quality apps for a low price is writing or note-taking. Google and its entire suite of tools (Docs, Sheets, Slides), available via web and mobile app, are extremely robust, sync easily to the cloud via Google Drive, and offer nearly the same number of features as traditional programs like Microsoft Word. In fact, this entire chapter was researched, written, and revised using Google Docs and Google Drive. In 2013, Apple began offerings Pages, Numbers, and Keynote to users who upgraded or purchased a mobile device running iOS 8. This elegant suite of tools, which was designed first as a mobile-only collection, allows you to create documents, edit, add formatting, and easily save to the cloud. While Microsoft has made its Office tools available via an app, a subscription is required in order to create new documents, provide major revisions, or use rich formatting. For those wanting to edit Office documents, Citrix ShareFile, QuickEdit, and WPS Office are all free apps that allows for creating, viewing, and editing documents. There are numerous other paid apps worth exploring, including Textilus and iA Writer.

## MANAGING PDFS

Mendeley is one of the most popular apps for managing PDFs. With free account creation, Mendeley can be synced across devices, gives 2 GB of free online storage, and lets you annotate and make notes. When linked with the web or desktop version, Mendeley becomes a powerful citation and bibliography tool as well. Qiqqa (Android only) is another popular tool that lets you manage PDFs, create annotation reports of all the notes you have made across multiple documents, and guides you through your literature based on identified themes. PDF Expert 5 (iOS only) lets you manage your PDFs, create freehand drawings and annotations, and even merge PDFs and manage pages within a PDF.

## CREATING BIBLIOGRAPHIES

Several well-known citation tools, such as Papers and EndNote, offer free apps but require a subscription to a desktop version in order to access or use content and do not contain integrated writing apps to create in-app citations. Instead, citations will mostly occur by using the copy-and-paste method. Most of the free citation apps are designed more with students in mind and are best for citing a few documents or websites in a single paper. For example, the free version of the EasyBib app is a citation generator that has over 7,000 citation styles available. Citations can be pulled from websites or by scanning the barcode of a book, and there are fifty-nine different publication formats supported (websites, books, articles, etc.). EasyBib, however, is not intuitive and gets cumbersome when one is dealing with more than a few references. BibMe and Citation Machine work much the same way, except for the barcode-scanning ability. By far the best free tool is for citation management is RefME. Although it does have a web interface, the app allows you to quickly search for citations in journals, books and websites or by scanning barcodes. It easily creates bibliographies that can be exported in a number of ways in over 7,500 citation styles.

## *Data Visualization*

Turning research data into a visual representation is a way to convey information in an entirely new way. Depending on the type of visuals required, there are multiple apps available. In order to move away from standard Excel pie charts, there are a host of tools that allow for the creation of beautiful charts, graphs, and infographics. Visme, Genial.ly, and Easel.ly all have basic models that allow for presentation, poster, and infographic creation. For those with some grant funding, Infogr.am is a monthly subscription–based web app that allows one to upload data sets, which it then uses to create interactive and custom infographics using hundreds of templates. ChartBlocks is a web-based app that allows for the creation of thirty free charts. You can copy and paste data or enter it manually and have the choice of five different chart types. The charts themselves are highly customizable in terms of color; can be downloaded as PS, SVG, and PNG files; and can be embedded into presentations or reports.

Publishers usually require at least 300 dpi for images to be included in publication and typically prefer image files in TIFF, EPS, or PDF format. To convert an image to a different dpi, researchers can use the web app Convert Town or download the free app for the open-source imaging software GIMP. To convert a file type to TIFF, there are numerous web-based products, including ConvertImage, Zamzar, and Sciweavers.

## *Publication and Dissemination*

### DESIGNING PRESENTATIONS

There are also a multitude of tools available for the creation of presentations using a mobile device. In terms of presentations, Google Slides is free, syncs with the Google Drive suite, and offers fairly robust formatting features as well as the ability to share presentations via URL and to present online. While the desktop version is more robust, Prezi Lite Editor is a free app that allows for the creation, editing, and online viewing of immersive presentations with a fairly wide array of themes. For librarians with a bit of funding or who have purchased an iPad recently, Keynote is popular with iOS users. It is very easy to use on a mobile device and allows for the creation and editing of visually appealing presentations using a large number of available templates. When presenting, Keynote has airplay support, which allows the presentation to be playing on the display device but for the notes and movement functions to be visible to the presenter on his or her device. For those researchers who have visual-heavy presentations, Haiku Deck is a great option. It requires a $5 USD monthly fee but has several nice features, including the ability to use your smartphone as the remote for controlling the presentation and access to a large collection of stock images and templates.

## CONFERENCE CALLS FOR PAPERS

Keeping track of calls for papers from conferences can be overwhelming. The low-cost Call For Papers pulls data from WikiCFP—which currently has one hundred CFPs in information science, sixty-four in digital libraries, and ninety-three in health informatics. Users select a favorites list that allows them to be notified when there are new calls in their area of research.

## SOCIAL MEDIA

As mentioned above, social media is one of the tools researchers already use in their research process. Twitter allows librarians to build up followers from around the world who are interested in the same research topics and then to tweet out new research projects or publications. Academia.edu is a social networking site for academics where researchers can post their CVs, upload preprints or other copies of their research, and connect with potential collaborators. The free app is not as robust as the website. However, it serves all basic functions.

## METRICS

Keeping on top of author- and article-level metrics is an important task for any researcher. In terms of alternative metrics, using the mobile-friendly site Impactstory, you can link your ORCID profile and will be provided with an in-depth analysis of your work, including mentions in Twitter, Facebook, Google+, and the news and information on your readability and global reach. For traditional metrics, accessing the mobile-friendly site Google Scholar and creating an online profile will allow you to view your $h$-index, i10-index, and citation counts.

## Archiving

There are several aspects of archiving that must be considered when discussing apps. The archiving and potential sharing of research data, particularly data from projects that were publically funded, is frequently a funding requirement. Unfortunately, there are no great options when it comes to data sharing and archiving that are available via app or mobile-friendly sites—likely the result of the large and numerous files and the often complex requirements involved in data archiving. Researchers will have to use standard websites and programs based on their geographic location, potential institutional programs, and funding requirements (if received) for the archiving of data.

In terms of the archiving of research papers or supplementary material, there are options depending on the type of material that needs to be archived. Dropbox and Google Drive both allow for uploading of documents and other file types, and researchers can publicly or privately share or post links to the material as they see fit. Many academic organizations also allow for the uploading of documents to institutional repositories (e.g., DSpace)—which unfortunately are not usually mobile-friendly. However, depending on institutional or granting agency requirements regarding access to data, as well as confidentiality and privacy concerns, researchers are advised to proceed with caution and consult local policies. Research data may need to be anonymized, and depending on the data management plan in place, there may be requirements surrounding retention and destruction. Data management is a rapidly emerging field within academia, and as yet there are no well-designed mobile-friendly products in place.

## Apps in Action

As mentioned above, there has been little work done on the use of apps to complete research, whether in the field of librarianship or elsewhere. Instead, the use of these tools is often "hidden" in the research process, such as the matter-of-fact way that collaborators will create a shared Google Drive folder or an Evernote notebook. It is usually only the apps or tools that are used in the completion of the Methods sections of papers that are highlighted simply because the authors must describe in detail how they completed the research. In some cases, the use of apps has become the norm—such as in the case of survey tools. SurveyMonkey, FluidSurveys, and Google Forms have become ubiquitous in librarianship—a field that loves a good survey. A quick scan through recent library research that used surveys will likely reveal that at least one of these tools was used. Other tools, such as Research Randomizer, do make appearances, particularly in health sciences research (Bothung, Fischer, Schiffer, Springer, & Wolfart, 2015; Orak et al., 2016). QuickVoice Recorder has been used in a number of studies, including one that looks at voice memory tasks (Lee, Sullivan, & Schneiders, 2015). Ultimately, however, the use of apps to complete research is not something that is frequently shared in the literature, just as most authors would not indicate that they used Microsoft Word to draft the text or Gmail to e-mail it to colleagues. If mobile apps and tools are going to continue to increase in popularity among library researchers, it seems likely that the sharing of these tools will have to occur largely through word-of-mouth and shared experiences. Knowing which of the millions of available apps is the one app that can help you to complete your research can save you untold amounts of time.

## Conclusion and Future Directions

This chapter has been an exploration of the mobile apps and tools that can help a librarian researcher complete a research project from start to finish. While some steps will be easier to complete on a mobile device, all the steps are, in fact, possible. As librarians have moved away from the stereotype of being hunched over card catalogs, so too must we move away from the idea that we must hunch over our desktop computers to perform our research. Mobile technology gives us the freedom to use a single device to conduct our research wherever suits us best—whether that is up in the stacks, out in public study areas, as part of clinical rounds, or embedded in the classroom. While the apps discussed here are up-to-date as of the spring of 2016, no doubt changes will occur as new tools appear and old ones disappear, and we will once again have to familiarize ourselves with new technologies. Keeping abreast of how our faculty members and students interact with one another and complete their own research will be an ongoing advantage in remaining vital to our communities and our institutions.

## References

Aiyegbayo, O. (2015). How and why academics do and do not use iPads for academic teaching? *British Journal of Educational Technology, 46*(6), 1324–1332.

Al-Aufi, A., & Fulton, C. (2015). Impact of social networking tools on scholarly communication: a cross-institutional study. *The Electronic Library, 33*(2), 224–241. doi:-doi:10.1108/EL-05-2013-0093.

American Dialect Society. (2010). "App" voted 2010 word of the year by the American Dialect Society. Retrieved from http://www.americandialect.org/American-Dialect-Society-2010-Word-of-the-Year-PRESS-RELEASE.PDF.

American Library Association. (2009). ALA's Core Competences of Librarianship Retrieved from http://www.ala.org/educationcareers/careers/corecomp/corecompetences.

Apple. (2013). App Store Sales Top $10 Billion in 2013. Retrieved from https://www.apple.com/uk/pr/library/2014/01/07App-Store-Sales-Top-10-Billion-in-2013.html.

Berg, S. A., Jacobs, H. L., & Cornwall, D. (2013). Academic librarians and research: a study of Canadian Library administrator perspectives. *College & Research Libraries, 74*(6), 560–572.

Bothung, C., Fischer, K., Schiffer, H., Springer, I., & Wolfart, S. (2015). Upper canine inclination influences the aesthetics of a smile. *Journal of Oral Rehabilitation, 42*(2), 144–152.

Bricker, J. B., Mull, K. E., Kientz, J. A., Vilardaga, R., Mercer, L. D., Akioka, K. J., & Heffner, J. L. (2014). Randomized, controlled pilot trial of a smartphone app for smoking cessation using acceptance and commitment therapy. *Drug & Alcohol Dependence, 143*, 87–94. doi:10.1016/j.drugalcdep.2014.07.006.

C-EBLIP. (2016). Centre for Evidence Based Library and Information Practice. Retrieved from http://library.usask.ca/ceblip/.

Calvi, L., & Cassella, M. (2013). Scholarship 2.0: analyzing scholars' use of Web 2.0 tools in research and teaching activity. *Liber Quarterly, 23*(2).

Canadian Association of Research Libraries. (2010). Core competencies for 21st Century CARL librarians. Retrieved from http://www.carl-abrc.ca/uploads/PDFs/core_comp_profile-e.PDF.

Canadian Association of Research Libraries. (2016). Librarians' Research Institute. Retrieved from http://www.carl-abrc.ca/lri.html.

Carson, P., Colosimo, A. L., Lake, M., & McMillan, B. (2014). A "partnership" for the professional development of librarian researchers. *Partnership: The Canadian Journal of Library and Information Practice and Research, 9*(2), 1–3.

Cohen, P. (2008 (July 14)). Apple: 10 million apps, 1 million iPhone 3Gs. Retrieved from http://www.macworld.com/article/1134484/appsphones.html.

Cohen, P. (2008 (Sept 9)). App Store downloads top 100 million, games a centerpiece. Retrieved from http://www.macworld.com/article/1135453/appstore.html.

Dilger, D. E. (2014 (July 10)). 75 billion downloads later, Apple celebrates the App Store's sixth anniversary. Retrieved from http://appleinsider.com/articles/14/07/10/apple-inc-reaches-sixth-anniversary-of-the-app-store.

Duncan, V., Kumaran, M., Lê, M.-L., & Murphy, S. (2013). Mobile Devices and their use in library professional practice: The health librarian and the iPad. *Journal of Electronic Resources Librarianship, 25*(3), 201–214.

Hall, L. W., & McBain, I. (2014). Practitioner research in an academic library: Evaluating the impact of a support group. *The Australian Library Journal, 63*(2), 129–143.

Haustein, S., Bowman, T. D., Holmberg, K., Peters, I., & Larivière, V. (2014). Astrophysicists on Twitter: An in-depth analysis of tweeting and scientific publication behavior. *Aslib Journal of Information Management, 66*(3), 279–296. doi:10.1108/AJIM-09-2013-0081.

Hennig, N. (2014a). *Apps for librarians: Using the best mobile technology to educate, create, and engage.* Santa Barbara, CA: Libraries Unlimited.

Hennig, N. (2014b). Evaluating Apps. *Library Technology Reports, 50*(8), 15–17. Retrieved from http://uml.idm.oclc.org:9080/login?url=http://search.ebscohost.com/login.aspx?direct=true&db=a9h&AN=99938278&site=ehost-live.

Hennig, N., & Nicholas, P. (2014). *Apps for Academics.* Publisher: Authors.

Ingraham, N. (2013 (Oct. 22)). Apple announces 1 million apps in the App Store, more than 1 billion songs played on iTunes radio. Retrieved from http://www.theverge.com/2013/10/22/4866302/apple-announces-1-million-apps-in-the-app-store.

Kalb, H., Bukvova, H., & Schoop, E. (2009). The digital researcher: Exploring the use of social software in the research process. *Sprouts: Working Papers on Information Systems, 9.*

Kennedy, M. R., & Brancolini, K. R. (2011). Academic librarian research: A survey of attitudes, involvement, and perceived capabilities. *College & Research Libraries*, crl-276.

Koufogiannakis, D., & Crumley, E. (2006). Research in librarianship: Issues to consider. *Library Hi Tech, 24*(3), 324–340. doi:doi:10.1108/07378830610692109.

Lee, H., Sullivan, S. J., & Schneiders, A. G. (2015). Does a standardised exercise protocol incorporating a cognitive task provoke postconcussion-like symptoms in healthy individuals? *Journal of Science and Medicine in Sport, 18*(3), 245–249.

Orak, M. M., Gumustas, S. A., Onay, T., Uludag, S., Bulut, G., & Boru, U. T. (2016). Comparison of postoperative pain after open and endoscopic carpal tunnel release: A randomized controlled study. *Indian Journal of Orthopaedics, 50*(1), 65.

Powell, R. R., Baker, L. M., & Mika, J. J. (2002). Library and information science practitioners and research. *Library & Information Science Research, 24*(1), 49–72.

Procter, R. N., Williams, R., & Stewart, J. (2010). *If you build it, will they come? How researchers perceive and use Web 2.0*. Retrieved from London UK: http://www.rin.ac.uk/system/files/attachments/web_2.0_screen.pdf.

Smith, B., Jacobs, M., & Lippincott, J. K. (2010). A mobile future for academic libraries. *Reference Services Review, 38*(2), 205–213.

Statista. (2016). Most popular Apple App Store categories in March 2016, by share of available apps. Retrieved from http://www.statista.com/statistics/270291/popular-categories-in-the-app-store/.

Strain, M. (2015 (Feb 13)). 1983 to today: A history of mobile apps. Retrieved from http://www.theguardian.com/media-network/2015/feb/13/history-mobile-apps-future-interactive-timeline.

Walker, H. (2010). Evaluation Rubric for Educational Apps. Retrieved from http://learninginhand.com/blog/evaluation-rubric-for-educational-apps.html.

Watson-Boone, R. (2000). Academic librarians as practitioner-researchers. *The Journal of Academic Librarianship, 26*(2), 85–93.

Woods, B. (2016 (Jan 20)). Google Play had twice as many app downloads as Apple's App Store in 2015. Retrieved from http://thenextweb.com/apps/2016/01/20/google-play-had-twice-as-many-app-downloads-as-apples-app-store-in-2015/#gref .

CHAPTER 10*

# A Novel Application
## Using Mobile Technology to Connect Physical and Virtual Reference Collections

*Hailie D. Posey*

## Introduction

Librarians are constantly considering the most effective ways to connect our users with content. This chapter outlines an innovative implementation of iPad kiosks to blur the lines between physical and virtual library collections. The kiosk presents the Theology Collections Portal, a web-based guide to electronic resources in theology available from Providence College's Phillips Memorial Library. The Theology Collections Portal content is presented to students and faculty through an iPad kiosk physically located within the library's theology collection (figure 10.1).

Providence College's Phillips Memorial Library + Commons began lending iPads to students, faculty, and staff in 2012. In addition to lending the devices, library staff dedicated time to learning about both task-based and subject-based mobile applications that would be of use to our community. A small group of library staff tested, discussed, and vetted a variety of apps that would be installed on the circulating iPads. Efforts were made to promote the use and discovery of various apps through thoughtful organization of the apps on the devices themselves, programming around applications, and the creation of an online research guide

---

* This work is licensed under a Creative Commons Attribution-NonCommercial 4.0 License, CC BY-NC (https://creativecommons.org/licenses/by-nc/4.0/).

designed to teach the community more about the apps. Despite these initiatives, assessment data from the iPad-lending program suggests that patrons borrowing the iPads use them primarily for accessing the Internet (Safari, Chrome, etc.) and social media (Facebook, Twitter, etc.) and consuming media (YouTube, Netflix, Pandora, Spotify, etc.) (DeCesare, Poser & Belloti, 2013, p. 33–36).

**FIGURE 10.1**
The Theology Collections Portal situated within the theology collection.

With this analysis in mind, library staff began to think of alternative ways to connect our patrons with useful, content-based mobile applications and library resources. Drawing on research focused on the Internet of Things and the integration of digital technologies with our daily lives, the Digital Publishing Services Coordinator suggested positioning iPad kiosks strategically within the library's physical book collection as a means of connecting patrons with the library's online resources related to a given subject area. Planning then began to bring a subject-based kiosk to life. Working collaboratively, the Digital Publishing Services Coordinator and the Commons Technology Specialist discussed how to image the iPad and how to manage the kiosk for optimal usability. The Digital Publishing Services Coordinator suggested that the kiosk pilot be aimed at students doing research in theology. Providence College is a Dominican Catholic institution, and students at Providence College are required to take coursework in the development of western civilization. Anecdotes from research librarians, as well as data from our research question–tracking system, LibStats, revealed a high frequency of research questions from theology students. This history of research inquiries, couples with the library's existing relationship with several theology faculty prompted the decision to develop pilot content for the kiosks with the aim of connecting theology scholars and students to the library's theology resources.

While information kiosks are present in many public spaces (retail stores, banks, airports, etc.), literature on information kiosks in libraries is relatively scant. Publications on the subject include an analysis of government information kiosks, library kiosks as OPAC terminals, a review of the role of health information kiosks (Wang & Shih, 2008), a discussion of an iPad vending machine kiosk at Drexel University ("Drexel U", 2015), and the use of OverDrive kiosks for browsing e-book collections at school and public libraries ("Kiosks made easy," 2014; Sun, 2015). One more robust review of the use of kiosks describes work done by a team of librarians at Texas Tech University to create an information kiosk that aims to provide "general information for frequently asked questions in a more efficient, creative, and interactive way" (Litsey et al., 2015, p. 31). Kiosk design at Texas Tech was guided by graphic design principles, including the explicit goal of being "adaptive, interactive, and usable" (Litsey et al., 2015, p. 33). The kiosk was built using a touch-screen Smart Board and programmed locally using PHP, JQuery, and Flash. Analysis of the information kiosk usage at Texas Tech has shown that the frequency of certain directional questions at the library has declined as a result of making the information readily available to patrons via the kiosk. The Texas Tech study, with its emphasis on design and usability, provides a good road map for considering a local kiosk implementation.

As part of our iPad initiative at the Phillips Memorial Library, we purchased a stand-alone iPad kiosk that housed an iPad 2 ("Standalone iPad Kiosk", 2016). The kiosk was placed in a high traffic area on the first floor of the library and was used as a quick-look-up library OPAC station. While the kiosk did get some usage, the library iPad team agreed that repurposing it in order to experiment with subject-based kiosk content was a better and more interesting use of the tool. Before the existing OPAC terminal kiosk was moved and reprogrammed, research was undertaken to plan for content options on the subject-based theology kiosk. A review of the literature in which librarians critically evaluated research guides proved critical to understanding strengths and weaknesses of guide creation and using that knowledge to inform the design of the theology collections kiosk. The body of writing on research guide effectiveness highlights concepts of usability and user experience that provide the foundation for choices around content and layout of the theology kiosk pilot project at Providence College.

# Research Guides and the User Experience

Research guides, also called subject guides or pathfinders, are web-based tools curated by librarians that provide students and researchers both access to and information about library resources. Librarians spend a great deal of time and energy creating these guides at both the discipline and course levels. Many libraries subscribe to content management system, which librarians use to create research guides, the most popular of which is Springshare's LibGuides system (Springshare, 2016). A recent study of ARL libraries found that 71 percent (of 101 responding libraries) used LibGuides as a CMS (Jackson & Stacy-Bastes, 2016, p. 222). Providence College also subscribes to LibGuides.

Much attention is paid to research guides in the library literature. In their recent analysis of the "Enduring Landscape of Online Subject Research Guides," Jackson and Stacy-Bates identify three primary themes of research guide–related scholarship present in the literature since 2002: guide content and arrangement, the use and discoverability of guides, and the promotion of guides (Jackson & Stacy-Bastes, 2016, p. 220–221). Notably absent from these discussions is research on usability and the user experience in relation to research guide use and creation (Jackson & Stacy-Bates, 2016, p. 228; Sinkinson, Alexander, Hicks & Kahn, p.63).

The literature on library research guides provides interesting observations about how librarians approach the research guide creation process. In his book *Modern Pathfinders: Creating Better Research Guides,* Jason Puckett observes that librarians tend to be text-centric and "completionist," meaning they believe it is "better to include too much information than too little" and that "a long list is better because it's more comprehensive and more likely to include the needed resource" (Puckett, 2015, p. 64). While this tendency to provide exhaustive information is immensely valuable in some situations, text-heavy and lengthy content is not often well suited to writing for the Web. In fact, in interviews with students regarding their perceptions of research guide usability at Concordia University

College of Alberta, Dana Ouellette found that the majority of the students were confused by unclear tab labels that included library jargon and that the majority of these students felt confused and discouraged when attempting to navigate a research guide (2011, p. 446–447). Beyond misunderstandings of language between research guide users and creators, several researchers have pointed to inherent differences between the mental models of research used by librarians and those used by students (Sinkinson et al, 2012; Reeb & Gibbons, 2004; Ouellette, 2011). Clearly, students and librarians have different needs and conceptualizations of research. While librarians "recognize that students approach research differently, research guides often reflect librarian mental models of research rather than replicating student preferences" (Sinkinson et al, 2012, p. 78). Librarians appear to want students to strive to emulate their model of research knowledge and thus create guides that present "the ideal" of information ecosystem understanding: an ideal that very few students will need or want to achieve as undergraduates. While librarians have a responsibility to teach students information literacy, this research suggests that librarians should consider a tiered approach to content creation in which the intended audience of a guide and that audience's information goals are given greater consideration. The needs of an undergraduate quickly looking for information in order to complete an assignment are very different from the needs of a graduate student or faculty member embarking on an in-depth research project.

Of the recent studies that account for user perceptions of research guides, a theme emerges around guide content specificity in relation to coursework. Reeb and Gibbons (2004) suggest that "library resources organized or delivered at the course level are more in line with how undergraduates do research" (p. 123). They suggest that undergraduate students' research mental models are more focused on coursework than on the scholarly discipline. This mental model, they contend, is not well suited to library subject guides that require an understanding of the discipline (Reeb and Gibbons, 2004, p. 126). Course-level guides that are highly customized to meet specific information needs would be much more useful and helpful to undergraduate students. To students who live in a digital environment laden with customization options, a general research guide appears impersonal and often unhelpful as it assists in the completion of no discrete, course-based project or assignment. Ouellette drives home this point by stating that "research has overwhelmingly shown that undergraduate students search for information in the easiest way possible to complete research quickly" (2011, p. 437). In order for research guides to add value for undergraduate users, they need to be customized, personalized, clear, and easy to use.

Because the kiosk project is focused on creating a mobile-friendly guide to the library's theology collections, a brief review of the work of Gerrit van Dyk (2015) on theology LibGuides is also warranted. Van Dyk examined theology and religion LibGuides at thirty-seven institutions. He found similarities in the structure and organization of the guides examined, but otherwise did not find a great deal of overlap. Tabs (or pages) on the theology LibGuides the author examined fell into the categories of "topical tabs, format tabs, religious tradition tabs, and help tabs" (van Dyk, 2015, p. 41). Only five of the thirty-seven guides contained religious tradition tabs. Van Dyk remarked that "this was surprising considering how many of the LibGuides were functionally combined Religious Studies/Theology guides, where information on religious traditions would be thought likely to find a place" (2015, p. 41). Of the 305 total databases listed on the guides examined, 143 were unique. Van Dyk states that "the resources and collections highlighted were so diverse that there were only a few resource types, databases in particular, where data overlapped enough to be of interest to theological librarians" (2015, p. 37). Eighteen of the 143 unique databases were used on four or more of the guides. Van Dyk's findings point back to the earlier assertion that librarians are completists and tend to take an everything-but-the-kitchen-sink approach. The extreme variety of database resources for a single discipline is surprising, and one must question the utility of presenting such a wide range of resources to students. Van Dyk concludes his discussion with a musing on whether or not the division of LibGuides by format is valuable to users or whether they would be better served by expanded topical tabs built to meet the needs of the local audience (van Dyk, 2015, 44).

Building on the aforementioned research, as the content and organization of the Theology Collections Portal developed, explicit consideration of the kiosk portal's target audience became key. At Providence College a large portion of the student body engages with theology resources. Undergraduate students are required to take four semesters of coursework in the development of western civilization (DWC). Many theology courses are cross-listed with DWC. This means that many undergraduates conduct research on theology topics early in their academic career. However, a majority of these students will not pursue theology as their main course of study. Non-majors, in particular, are likely to exhibit the mental models of research outlined by Reeb and Gibbons (2004), Ouellette (2011), and Sinkinson and colleagues (2012), that is, that they will seek to locate information quickly in order to complete coursework but are less interested in gaining deep knowledge of the conventions of theology scholarship. These students, novice undergraduate researchers working in theology and religion, are the target audience for the Theology Collections Portal. Its aim is to connect such students with the library's theology collections in a way that is quick and comprehensible.

With a stated audience in mind, design of the Theology Collections Portal project began with a focus on usability and the user experience. In line with the relevant literature, throughout the design process the aim of the kiosk portal was to meet users where they are (Reeb and Gibbons, 2004, p. 129), create guides that align with student mental models of research and actual user needs (Sinkinson et al, 2012, p. 64), and rise to student expectations (Quintel, 2016, p. 8). Of course, only extensive usability testing and continued conversation with students and faculty around the effectiveness of the kiosk in meeting these needs will determine if this user-centered approach was, in fact, helpful. Nevertheless, recommendations put forth by those researchers who have done work on subject guide usability were carefully considered and implemented in order to optimize the user experience of the Theology Collections Portal. These practices included engaging users with interesting visuals; using simple, easy-to-understand language; avoiding jargon; and presenting clear pathways through content, making navigation seamless.

## Building the Theology Collections Portal in Scalar

The theology kiosk was initially conceived of as an iPad kiosk that would be loaded with subject-specific content and organized using web clips. After completing the investigation around research guides and their usability, it became clear that the presentation, navigability, and general usability of the kiosk were of the utmost importance if it was to be an effective tool. The author considered a variety of content management systems that might allow users to follow customized paths into the kiosk based on their particular research needs. Fortuitously, the author and the Head of Digital Publishing Services and Cataloging happened to be collaborating with a local cultural heritage institution around the creation of an open-access e-textbook. Among the tools being considered for the e-text project was Scalar. Scalar is a digital publishing tool created by the Alliance for Networked Visual Culture (ANVC) led by Tara McPherson at the University of Southern California in Los Angeles. The ANVC "aims to close the gap between digital visual archives and scholarly publication by enabling scholars to work more organically with archival materials, creating interpretive pathways through materials and enabling new forms of analysis" (McPherson, 2010, p. 6). In support of this mission, the ANVC created Scalar, which was released in beta in 2011 and has recently been released as Scalar 2 (The Alliance for Networking Visual Culture , 2016). Scalar is a content management system that seeks to provide scholars with the ability to create publications that include a variety of media. Users can embed media alongside text to easily create a digital "book" or website. Pages are created using a variety of flexible templates, and content can be viewed in a variety of ways, including "media-centric views, text-centric views, graph views, grid views, etc." (McPherson, 2010, p. 7). A signature design element in Scalar is the ability to create multiple narrative paths through a work. Scalar also allows for extensive annotation of both text and media by both authors and readers. Additionally, Scalar presents built-in visualization tools, which allow creators to explore and adjust the relationships between content in different ways.

Only two studies in the library and information studies literature make reference to Scalar, and both do so in the context of library support for the digital humanities. Anita Say Chan and Harriet Green (2014) examine Scalar as a tool for fostering digital literacy in the humanities classroom. They suggest that encouraging the use of Scalar and other new digital publishing tools and platforms as part of the humanities curriculum fosters "collaborative student engagement" and "encourage[s] playful student tinkering" (Chan & Green, 2014, p. 1). Their findings, based on case studies of two media and cinema studies courses in which a librarian partnered with a teaching faculty member to promote digital publishing tools, suggest that while students are avid consumers of digital content, they have less experience creating such content. They point to research that cautions against seeing "digital natives" as "independently able to generate the adequate literacies necessary to manage the complexity of networked lives" (Chan & Green, 2014, p. 18). Exposure to a variety of digital publishing tools, even those that are "under-tested" in the pedagogical sphere, "empowers students to produce interactive scholarship that develops their digital literacies by collating and evaluating research sources, synthesizing information, and designing new scholarly works" (Chan & Green, 2014, p. 18–19)

At the University of Illinois at Urbana–Champaign, librarians partnered with the platform's developers and began marketing Scalar to their user community as a digital publishing tool (Tracy, 2016). Tracy conducted a survey of twenty Scalar users (primarily faculty and graduate students) as well as eight interviews with Scalar users (seven faculty and one staff member). Survey and interview respondents provided several reasons Scalar appealed to them for research and teaching, among them the ability to incorporate multimedia and images, the ability to create multiple paths through a set of research objects, the desire to have students engage with technology in the classroom, and enabling complex annotation and conversation around digital items (Tracy, 2016, p. 170). Tracy's study underlines some usability issues within Scalar, including snags in the media upload process and difficulty using the annotation tools (these issues seem to have been addressed with the release Scalar 2). More importantly, Tracy's study echoes that of Chan and Green in pointing to the fact that some respondents (and their students) have not had the opportunity to develop the mental models necessary for teaching and learning in the digital environment. Tracy points to the fact that very few survey respondents or interviewees made use of Scalar's organizational flexibility, not making use of multiple paths through their content, but instead following a linear structure of a series of pages.

While no formal writing has been published around the use of Scalar to create research guide–like content, Scalar's functionality, specifically its nonhierarchical organization and promotion of multiple narratives, affords the opportunity to build an interactive tool in which users have greater control over their research experience. The relevant literature having been investigated, work began to create the Theology Collections Portal using Scalar (Theology Collections Portal, 2016; Posey, 2016). Content within the kiosk portal has been divided in to five major sections. The first three of these sections allow users to engage with and explore library resources; the final two provide information on getting further help from library staff and allow users to provide feedback on the kiosk.

## *Find Scholarly Sources for a Paper*

The first major path in the Theology Collections Portal is designed to connect users to scholarly content, both physical and virtual. Of the options provided to users, this path is the most closely related to traditional research guide structure in that it is divided by format. Users have the option to find articles in a theology database, find a specific journal, or find a book or e-book. The page presenting theology databases links to ATLA Religion Database, Catholic Periodical and Literature Index, and ATLA Historical Monographs Series 1 and 2 (Posey, 2016, p. 5). Users can directly access these databases and complete their search on the kiosk. Some users may want to access these databases on their own machine or device. Direct e-mailing of links from the kiosk is not possible because of the locked-down nature of Kiosk Pro App, the tool used to manage the iPad within the kiosk (more on Kiosk Pro App below). In order to make kiosk content available to patrons who may seek to view it on their own devices, cards containing the Portal URL and a QR code linking to it are available at the kiosk.

The page presenting specific journals links to the library's Publication Finder tool and explains how to use it to search for a journal by name (Posey, 2016, p. 6). It includes a couple of images of the Publication Finder and reiterates that one would use this tool to find and browse a specific journal or to find an article based on a citation. Patrons are then guided to a page instructing them on finding a book or e-book (Posey, 2016, p. 7). The page describes the search process and links to the library catalog. It shows students where to enter their keyword search term and briefly explains how to utilize the facets or filters to refine a search. It explains how to order books from other libraries within the HELIN Consortium and how to access e-books from within a catalog search.

## *Explore Theology Topics*

Currently, in this first iteration, the theology topics section of the kiosk contains three primary paths: Major Religions, Thomas Aquinas, and Catholicism and Catholic Social Thought (Posey, 2016, p. 8). These areas were selected through an analysis of research questions fielded by librarians over the course of the last five years. Research questions are documented using a system called LibStats. The author ran a search on questions containing the word *theology* or *religion* over the last five years. These questions were then coded to determine recurring questions and content inquiry themes (figure 10.2).

**FIGURE 10.2**
Theology-related research questions by category.

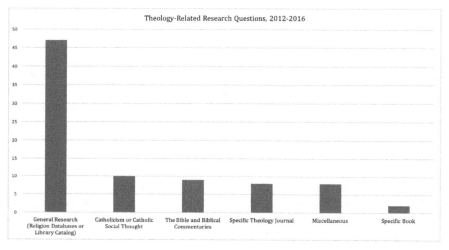

This section of the Theology Collections Portal could be significantly expanded through partnerships with theology faculty to contain pages for specific courses or specific assignments.

The Major Religions path begins with links to electronic reference materials on world religions including several dictionaries of world religion and an encyclopedia of global religion. Users are then able to select from a list of major religions: Judaism, Islam, Christianity, Buddhism, or Hinduism (Posey, 2016, p. 9). Each of these topic pages presents a variety of content in several formats. Preference was given to resources available electronically, but information about print resources is also included when the print resources better serve the research need. One example illustrates the general layout and presentation of the Major Religions pages. The Islam page (Posey, 2016, p. 10) presents the following content: a link to an electronic version of the Qu'ran; links to several digital reference works, call numbers for print reference works, the ability to browse journals in Islamic Studies using BrowZine Web,* links to a few relevant e-books, and a link to a precompleted search in the Encore catalog system presenting over 1,000 e-books on the subject of Islam (Posey, 2016, p. 12).

---

* BrowZine is both a mobile app and a website created by Third Iron, Inc., with the aim of presenting a library's journal holdings in a browse-able fashion. For more on BrowZine see http://thirdiron.com.

An additional precompleted search link providing a list of over 1,000 print books available at the library at Providence College is also listed, with a note about their location in the physical stacks (in most cases, directly behind the kiosk).

The topic pages related to Thomas Aquinas and Catholicism and Catholic Social Thought are similarly organized. Each page links to databases, as relevant; to specific e-books and to precompleted e-book catalog search; and to a list of several print books accessible near the kiosk. A major benefit of these topic pages is that they provide information about the topic in a variety of formats all in one place without the student having to navigate to discrete sections of the library website.

## *Find Bibles and Biblical Commentary*

Given the curriculum at Providence College, questions related to the Bible were extremely prevalent. Many questions also demonstrated that students sought biblical commentary but had trouble accessing it through the library catalog. Clear instructions about Bibles and biblical commentary are absent from the library's theology research guide (Providence College, 2016). The page dedicated to presenting this content could serve as a useful starting point for students in a variety of courses. As mentioned previously, one of the great benefits of Scalar is that each page or item has a unique URL. In this case (as is the case of some of the other topic pages), in addition to accessing the content from the kiosk in the library, faculty could provide direct links to this content in their syllabus or learning management system and students would have a significantly easier time accessing the resources.

The path dedicated to Bibles and biblical commentary contains four pages (Posey, 2016, p. 13). The first presents information on finding Bibles. It links to a digital edition of the King James version of the Bible from Oxford University Press and outlines where print Bibles are located in the theology collection directly behind the kiosk. The second page provides students with an explanation of three important research tools related to the Bible: biblical commentaries, biblical dictionaries, and biblical concordances (Posey, 2016, p. 14). Each kind of tool is defined, and links to electronic versions of the tools are provided in addition to call numbers for print resources. Users also have the option to browse biblical studies journals using BrowZine Web (see footnote). The Bibles path concludes with links to the Old and New Testament Abstracts databases.

## *Get Help and Provide Feedback*

The final two paths accessible from the Welcome page of the Theology Collections Portal provide users with the tools to get further help and provide feedback. The help page presents a stand-alone graphic reiterating that the library is glad to help with any research questions and provides contact options including text, e-mail, telephone, and face-to-face (Posey, 2016, p. 15). Making use of Scalar's functionality allowing content to be reused and multiple relationships to be built allows the graphic to be interspersed throughout the kiosk's content in an attempt to provide help at the point of need for a user browsing the Theology Collections Portal. The final page is a simple assessment tool. This page presents an embedded Google Form seeking feedback from users of the portal (Posey, 2016, p. 16). It seeks the following information: user status (undergraduate, graduate, faculty, etc.), affiliation with Providence College, whether the user found the kiosk useful, primary reason for using the kiosk, whether the user found it easy to navigate, and a couple of open-ended questions related to what users like about the kiosk and what they would like to see improved.

In addition to the collection of data from users of the kiosk (and recognizing that many users will not voluntarily complete the feedback survey), Google Analytics has also been installed on the Scalar site containing the Theology Collections Portal content. Google Analytics data has not yet provided much insight into kiosk use as the kiosk pilot launch occurred recently, but this data will be extremely valuable as future planning is undertaken.

## *Visualizations and Search*

One of the most interesting tools built into Scalar is the ability to visualize relationships within a Scalar book. Users can select the visualizations menu and choose from a variety of visualization options. For example, the Connections visualization creates an interaction map of all the content in the Scalar book in force-directed format (figure 10.3).

### FIGURE 10.3
Visualization of all content and relationships in the Theology Collections Portal in force-directed format.

**Visualization**

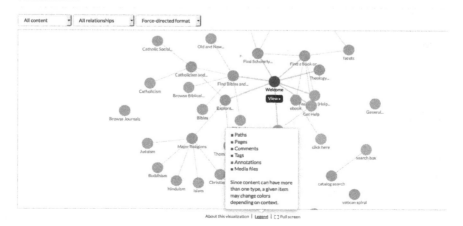

Not only does the visualization provide insight into the organizational structure and relationships built into the content, it also serves as an alternative navigation tool for those users who may not be more interested in exploring the Theology Collections Portal in a nonlinear fashion. An interesting exercise would involve replicating the traditional research guide in a Scalar book in order to visualize its structure and contrasting that with the structure and relationships built in the Theology Collections Portal.

Scalar also enables users to search for content (Posey, 2016, p. 18). A student could come to the kiosk looking for a certain kind of information and enter that term to connect with any page or other media type in which it is contained.

## *Kiosk Pro App*

While the content presented on the Theology Collections Portal is created in Scalar, the kiosk itself is managed using Kiosk Pro Plus App, version 7.2 (Kiosk Pro, 2016). The Kiosk Pro App provides a simple but robust management tool for various kinds of information kiosks. The iPad is set to keep the Kiosk Pro App open at all times. Kiosk Pro presents web content while preventing user access to the iPad home screen and settings, essentially locking down the iPad to the specified web homepage and permitted web domains. In order to program the Theology Collections Portal, the following settings were enabled:

- Homepage set to http://scalar.usc.edu/works/theology-collections/index.
- Allowed domains: Set the kiosk to allow domains pointed to by Scalar, for example, scalar.usc.edu, library.uri.edu, *.helin.uri.edu, *.eblib.com, encore.uri.edu, ebsco.com, browzine.com, and docs.gooogle.com.
- Restricted domains: Denied access to certain internal institutional domains that might pose a security risk were a guest to access the kiosk (e.g., the internal website).
- Idle time limit set to ninety seconds. Browsing time limit set to sixty minutes. If either of these thresholds is reached, the kiosk homepage is reloaded.

Kiosk Pro App has a variety of more advanced functionality, including the ability to use custom JavaScript to manage kiosk behaviors. Users can use the in-app editing tools or manage the kiosk from an external XML file living on a server. Given that the Theology Collections Portal content has been wrapped in a Scalar book, management of the kiosk has proven fairly simple.

## *Faculty Feedback*

After completing the initial Theology Collections Portal prototype in Scalar and testing it on the iPad kiosk, the author met with a member of the theology department in early April 2016 to share the work and discuss future directions. The theology faculty member was enthusiastic about the possibilities created by the kiosk. He mentioned the possibility of doing a demo to specific classes. He also offered to share basic information about the kiosk with theology faculty colleagues at an upcoming department meeting (May 2016) and welcomed the author to attend a meeting and present the kiosk in more detail the following semester. Partnering further with theology faculty to both promote and refine the kiosk will be crucial to its success. Through collaborations with faculty, the kiosk will make a broader impact on students. Students in theology courses who make use of the kiosk will also be able to provide feedback on its usability.

## Strengths, Shortcomings, and Future Directions for the Theology Collections Portal

The Theology Collections Portal guides users, specifically undergraduate students, through the library's digital collections in an engaging and informative way that encourages their active participation in the research process while simultaneously simplifying the research process to meet their specific needs. Visualization tools built into Scalar demonstrate a web of relationships and allow for the navigation of resources in new and interesting ways. Scalar is also highly customizable, and content creation is easy. This will facilitate further development of kiosk content designed to serve specific student needs based on course-related research needs.

An important consideration to which further attention must be paid is the accessibility of kiosk content for individuals with disabilities. Text on the kiosk is small and will not be suitable for anyone with a visual impairment. The text size is a result of Scalar's default styling. Custom styling could allow for increased text size, but that would obviously affect the amount of content on each page. While Kiosk Pro App does provide for the option to allow pinch and zoom functionality, Scalar does not. Other shortcomings of the kiosk as it currently exists relate to content and usability. These are areas that will be improved upon over time as more users interact with the tool and more data can be captured about usage.

The Theology Collections Portal kiosk project is in its infancy. If it is adopted as a major initiative of the Phillips Memorial Library, there are several directions the project could take and several areas that will require further study and refinement. These future directions include the following:

- Further outreach and collaboration with the theology faculty to maximize kiosk usability and refine content options, especially the creation of course-based topic pages.

- Partnering with existing library groups to collaboratively work on marketing and promotion of the kiosk.
- Collaboration with the library's Research and Education Department to determine how the kiosk can serve as a complement to our LibGuides or how the two tools might reference one another.
- Library-wide discussions around scaling and sustaining this project to include additional disciplines. This will necessitate good project management skills, especially in the area of assigning responsibility for kiosk content creation and maintenance.
- Determination of whether and how the creation of additional kiosks fits into the library's larger plan for marketing electronic resources.
- Further thought around the strengths and limitations of Scalar as a content management system for kiosk content.
- Performance of student-centered usability testing and focus groups to ascertain the most and least effective elements of the kiosk navigation, content, and design.
- Refinement of kiosk assessment tools and ongoing review of patron feedback around the kiosk to fuel continued improvement of the tool.
- Consideration of whether the addition of multimedia content, specifically video tutorials, could enhance usability of the kiosk.

Exploring digital research materials presented through a kiosk is more engaging for users than browsing a traditional research guide. Users must actively engage with the content in order to make their way through the kiosk content. Touch-based interaction may enhance the user's feeling of ownership over the research process, and the tactile experience of interacting with the kiosk may prove memorable and enhance the user's recall of certain elements. There is an intimacy and familiarity that comes from interacting with a mobile device that many patrons may find familiar and favorable when it comes to conducting library research.

The Theology Collections Portal is one example of how mobile technologies can aid in the library's goal of connecting patrons to digital resources. More broadly, the creation and presentation of the kiosk raise important issues around meeting our patrons at the point of need and facilitating coursework based on student mental models of the research process. The kiosk portal is an example of a library-created tool that engages mobile, responsive, user-oriented, design-driven thinking to better connect patrons with electronic resources while facilitating the research process. It presents one possible step toward a model of user-centered tool creation by librarians that is geared toward continued enhancement and refinement of the tool based on assessment and usability studies. Mobile technologies provide fertile ground for these kinds of innovative tools because of their ubiquity and the level of comfort most undergraduate students have with such technologies. It is critical that we put our users first and create tools that help them navigate the increasingly complex digital information ecosystem that our collections represent.

# References

Aegard, J. (2010). Library kiosks. *Computers in Libraries, 30*(8), 16–20. Retrieved from http://search.ebscohost.com/login.aspx?direct=true&db=lxh&AN=54005273&site=ehost-live.

Alliance for Networking Visual Culture, The. (2016). Scalar. Retrieved from http://scalar.usc.edu/.

Burdick, A., & Willis, H. (2011). Digital learning, digital scholarship and design thinking. *Design Studies, 32*(6), 546–56. doi:10.1016/j.destud.2011.07.005.

Chan, A. S., & Green, H. (2014). Practicing collaborative digital pedagogy to foster digital literacies in humanities classrooms. *Educause Review.* http://er.educause.edu/articles/2014/10/practicing-collaborative-digital-pedagogy-to-foster-digital-literacies-in-humanities-classrooms.

DeCesare, J., Posey, H., & Bellotti, C. (2013). Lending iPads 101: Steps to loan from your library. Presentation at NERCOMP Annual Conference, Providence, RI, March 11–13, 2014. https://works.bepress.com/hailie_posey/7.

"Drexel U introduces ipad vending machine."(2015). *Advanced Technology Libraries, 44*(5), 2–3. Retrieved from http://search.ebscohost.com/login.aspx?direct=true&db=lxh&AN=102648309&site=ehost-live.

Enis, M. (2014). Queens tests job app kiosks. *Library Journal, 139*(9), 13–13. Retrieved from http://search.ebscohost.com/login.aspx?direct=true&db=lxh&AN=95951685&site=ehost-live.

Jackson, R., & Pellack, L. J. (2004). Internet subject guides in academic libraries: An analysis of contents, practices, and opinions. *Reference & User Services Quarterly, 43*(4), 319–27. Retrieved from http://www.jstor.org/stable/20864244.

Jackson, R., & Stacy-Bates, Kristie K. (2016). The Enduring Landscape of Online Subject Research Guides. *Reference & User Services Quarterly, 53*(3), 219–29.

Joshi, A., & Trout, K. (2014). The role of health information kiosks in diverse settings: A systematic review. *Health Information & Libraries Journal, 31*(4), 254–73. doi:10.1111/hir.12081.

Kim, B. (2013). Responsive web design, discoverability, and mobile challenge. *Library Technology Reports, 49*(6), 29–30. Retrieved from http://search.ebscohost.com/login.aspx?direct=true&db=lxh&AN=90405356&site=ehost-live.

Kiosks made easy. (2014). *American Libraries, 45*(7), 42–43. Retrieved from http://search.ebscohost.com/login.aspx?direct=true&db=lxh&AN=97342266&site=ehost-live.

Kiosk Pro. (2016). Kiosk Pro App. Retrieved from http://www.kioskproapp.com/.

Litsey, R., Hidalgo, S., Daniel, K., Barnett, J., Kim, A., Jones, S., et al. (2015). Interactive kiosk at the Texas Tech University libraries. *Journal of Access Services, 12*(1), 31–41. doi:10.1080/15367967.2015.1020381.

McPherson, T. (2014). Designing for difference. *Differences: A Journal of Feminist Cultural Studies, 25*(1), 178–88. doi:10.1215/10407391-2420039.

McPherson, T. (2010). Scaling vectors: Thoughts on the future of scholarly communication.

*Journal of Electronic Publishing, 13*(2), 9. doi:10.3998/3336451.0013.208.

Mitchell, E. T. (2016). Library linked data: Early activity and development. *Library Technology Reports, 52*(1), 5–33. Retrieved from http://search.ebscohost.com/login.aspx?direct=true&db=lxh&AN=111864256&site=ehost-live.

Ouellette, D. (2011). Subject guides in academic libraries: A user-centred study of uses and perceptions. *Canadian Journal Of Information & Library Sciences, 35*(4), 436–51.

Posey, H. (2016). Theology collections portal: first iteration 2016. http://digitalcommons.providence.edu/facstaff_pubs/47.

Providence College, Phillips Memorial Library. (2016). Theology Research Guide. Retrieved from http://providence.libguides.com/theology.

Puckett, J. (2015). *Modern pathfinders: Creating better research guides.* Chicago, IL: Association of College and Research Libraries.

Quintel, D. F. (2016). LibGuides and usability: What our users want. *Computers in Libraries, 36*(1), 4–8. Retrieved from http://search.ebscohost.com/login.aspx?direct=true&db=a9h&AN=112316565&site=ehost-live.

Reeb, B. & Gibbons, S. (2004). Students, librarians, and subject guides: Improving a poor rate of return. *Portal: Libraries and the Academy. the Johns Hopkins University Press, 4*(1), 123–30.

Rosen, J. (2014). Islandport tests kiosks. *Publishers Weekly, 261*(53), 9. Retrieved from http://search.ebscohost.com/login.aspx?direct=true&db=lxh&AN=100084891&site=ehost-live.

Sandnes, F., Tan, T., Johansen, A., Sulic, E., Vesterhus, E., & Iversen, E. (2012). Making touch-based kiosks accessible to blind users through simple gestures. *Universal Access in the Information Society, 11*(4), 421–31. doi:10.1007/s10209-011-0258-4.

Sinkinson, C. & Alexander, S. & Hicks, A. & Kahn, M. (2012). Guiding design: Exposing librarian and student mental models of research guides. *Portal: Libraries and the Academy, 12*(1), 63–84.

Springshare. (2016). LibGuides. Retrieved from http://www.springshare.com/libguides/

Standalone iPad Kiosk. (2016) Retrieved from https://www.ipadkiosks.com/collections/ios-kiosks/products/standalone-ipad-kiosk.

Sun, C. (2015). Industry: School, public library partner on E-kiosk. *Library Journal, 140*(3), 19. Retrieved from http://search.ebscohost.com/login.aspx?direct=true&db=lxh&AN=100872988&site=ehost-live.

Theology Collections Portal. (2016). Retrieved from http://scalar.usc.edu/works/theology-collections/index.

Tracy, D. G. (2016). Assessing digital humanities tools: Use of scalar at a research university. *Portal: Libraries & the Academy, 16*(1), 163–89. Retrieved from http://search.ebscohost.com/login.aspx?direct=true&db=lxh&AN=113232204&site=ehost-live.

van Dyk, G. (2015). Finding Religion: An Analysis of Theology LibGuides. *Theological Librarianship, 8*(2), 37–45. Retrieved from http://search.ebscohost.com/login.aspx?direct=true&db=lxh&AN=110873368&site=ehost-live.

Vileno, L. (2007). From paper to electronic, the evolution of pathfinders a review of the literature. *Reference Services Review, 35*(3), 434–51. Retrieved from http://search.ebscohost.com/login.aspx?direct=true&db=eoah&AN=34597760&site=pfi-live.

Yan, Q. L., & Briggs, S. (2015). A library in the palm of your hand: Mobile services in top 100 university libraries. *Information Technology & Libraries, 34*(2), 133–48. doi:10.6017/ital.v34i2.5650.

Yi-Shun, W., & Ying-Wei, S. (2009). Why do people use information kiosks? A validation of the Unified Theory of Acceptance and Use of Technology. *Government Information Quarterly, 26*(1), 158–165. doi:10.1016/j.giq.2008.07.001.

CHAPTER 11*

# Adding Apps to Our Collections
## A Pilot Project

*Willie Miller, Yoo Young Lee, and Caitlin Pike*

## Introduction

In 1534, William Tyndale published the first Christian Bible translated directly from the original Hebrew and Greek texts. Previously, the most popular Bible was the Wycliffe Bible, which was translated from a Latin translation into Middle English. Tyndale's Bible was written in the commonly used, easily read language. When it was published, it was condemned in England and by the Catholic Church, essentially because it was too accessible to people. An English Bible was thought to reduce the power of the Church because the Church is where people heard the scripture as mediated by priests. Letting people read the Bible on their own was considered too dangerous because people could make their own decisions and come to their own interpretations and maybe make mistakes. In some ways, this sounds similar to the reaction of libraries to the Internet about twenty years ago. The gatekeepers of knowledge reacted to the gate being opened. Libraries were afraid of what the Internet might do to the nature of their relationships with users, and it is true that our relationship has changed, but for the better. A similar change in technology is happening in the area of mobile technology that will also have a profound impact on society in general, and libraries in particular.

---

* This work is licensed under a Creative Commons Attribution 4.0 License, CC BY (https://creativecommons.org/licenses/by/4.0/).

The mobile world is a fact, and mobile devices are ubiquitous. Due to their size and relative cost, smartphones reduce the digital divide for low-income populations and those living in developing countries. In fact, many people are now more proficient using a mobile device than they are using a traditional computer. This is likely why Microsoft is in the process of developing one user interface across all devices (Spence, 2016)An increasing number of Americans are "smartphone-dependent," using their smartphone as the primary method for Internet access (Smith, 2015). Furthermore, anecdotal evidence suggests that they are using apps more than books. A large number of American smartphone users report using their devices to find information about health conditions, to use educational resources, and to access new (Smith, 2015).

Though the popular stereotype portrays a library as a warehouse of books and a place with computers, we know that we are far more than that. Libraries exist to provide access to information, to facilitate the exchange of ideas, and to break down barriers that limit achievement and innovation. To fulfill this purpose, it is time that libraries seriously begin to include mobile applications as a part of our collections. Strict allegiance to older, traditional formats for information can stifle our passion. The book has been more or less a static device for several centuries. Apps help create new connections by linking to other content, to audio and video, to engaging images, and to the wider world of information.

## Literature Review

A review of the literature shows that, as a profession, academic librarians have been trying to figure all of this out since the Apple iPad was released in 2009. Many of the early case studies of iPads focused on their use as an alternative to computers for reference service (Tao, McCarthy, Krieger & Webb, 2009; Lotts & Graves, 2011; McCabe & MacDonald, 2011; Maloney & Wells, 2012) or for services related to student learning through iPad circulation programs Shurtz, Halling, & McKay, 2011; Capdarest-Arest, 2013; Massis, 2013). Librarians at Southern Illinois University–Carbondale developed a roving reference program, which had librarians borrowing iPads from a limited collection for circulation. This program led the authors to value iPads as an "ideal" tool for reference services and to recommend that each librarian should have one (Lotts & Graves, 2011). Medical libraries show focused interest in circulating iPads as educational tools (Capdarest-Arest, 2013; Duncan, et al., 2013; Gillum & Chiplock, 2014). Academic and medical librarians clearly see the value in mobile tablets; however, relatively few studies concentrate on circulation services and information literacy instruction sessions.

In the medical and health sciences fields, mobile technology is already vitality important and projected to increase in use (Barker & Knapp, 2014; Deloitte Center for Health Solutions, 2012; Payne, Wharrad, & Watts, 2012). Practitioners and clinicians regularly use mobile devices to streamline reporting and paperwork and to update records. Moreover, many health-related apps outperform traditional material formats for accessing health information. It is important for health sciences students to have access to the necessary technology and devices they will be expected to use in the field. However, many of the best medical and health science apps are prohibitively expensive, require institutional subscriptions, or involve additional technology—all of which widens the digital divide.

Some faculty at IUPUI (Indiana University–Purdue University Indianapolis) explored the use of iPads for instruction soon after the devices were first released by Apple (Rossing, Miller, Cecil & Stamper, 2012). At that time, iPads were novel. This group, among them the first-named author of this chapter, sought to discover the possible educational advantages of the mobile devices. Over the three-year period of the study, the group used iPads in more than twenty courses. Deploying the devices in class for active learning activities, the group surveyed students on their perceptions of learning and engagement at the end of the semester. Students reported confidence in the iPad in aiding them to learn course content, applying content to solve problems, approaching ideas in new ways, and enhancing their learning (Rossing et al., 2012, p. 9).

In order to meet the growing demands for mobile resources in the health sciences fields, liaison librarians Willie Miller (Informatics and Computing), Yoo Young Lee (Health and Rehabilitation Sciences), and Caitlin Pike (Nursing), together with library staff, developed a pilot app-purchasing and mobile device-lending program at IUPUI. This pilot program made these resources accessible for students and faculty during information literacy instruction sessions. The objective was to demonstrate the value of mobile content and the viability of lending mobile devices at our institution. The pilot was devised for the following reasons:

1. The authors believe that mobile applications provide valuable information for users.
2. There is a need for this technology in the IUPUI community. The IUPUI campus did not have device checkouts from the library or other units, even though students in certain degree programs were required to use tablet devices.
3. Some apps are more engaging than some books or journals because of their multimodal content (words, images, and videos) and functionality—you can complete processes in apps digitally and through voice recognition in some instances.

Several reports indicate that most college students have mobile devices(Belardi, 2015; Poll, 2015), and the authors believe that this will change the way in which students access information in the health sciences fields and in other disciplines. It is central to the mission of an academic library to provide resources in any format for students' academic success.

## Project Background

The IUPUI University Library is a large research library on an urban campus, which consists of 30,000 undergraduate and graduate students and is located in Indianapolis, Indiana, USA. The library maintains a robust liaison librarian program in which each librarian is subject specialist, collection manager, and instruction librarian to the department or school to which he or she is assigned. This model of academic librarianship allows for a high level of professional autonomy and collaboration with faculty.

Since the library has a reputation as a leader in the application of technology to library practice; the library is given IT support from the Client Support Team (CST) as well as from campus University Information Technology Services (UITS). CST and UITS determined how to best share the iPads. CST is in charge of technical support such as maintenance and hardware upgrades at the library.

This project was funded from a Library Services and Technology Act (LSTA) grant to purchase thirty-one Apple iPad Minis (second generation) and thirty-one OtterBox Defender cases to protect the devices from physical damage. The library also purchased a Bretford Mobility Cart to securely store, charge, and sync the iPads. Additionally, the library has funded the cost of the health sciences apps and any institutional subscriptions that they require. The maintenance cost of the equipment has been folded into the annual budget for technology; however, the authors feel that this technology and similar programs should be a part of future library services and collection development budgets.

Even though the authors consider app acquisition to be collections work, the purchase process is different from that of traditional collection acquisition. At IUPUI University Library, the acquisitions team handles all of our collection development, including the processing of books, journals, and database subscriptions. However, for apps purchases, the acquisition team is not currently involved. First, the business manager/fiscal officer enrolled our organization in the Apple Volume Purchase Program (VPP) for Education. This program allows us to purchase apps in volume and distribute them through our iPads. Once librarians select apps to be added to iPads, CST purchases the apps and makes them available on the devices through MAM remote app updating.

## App Selection

The selection of apps for the iPads was determined by the unique needs of each liaison librarian's department or school. The schools of Nursing, Health and Rehabilitation Sciences, and Informatics and Computing have some overlap in subject areas, but each has a specific focus that we wanted to highlight. In choosing the apps, the authors used iMedicalApps, a physician-reviewed mobile health site, and consulted with several medical librarian colleagues who had already started to develop iPad programs at their own libraries. One of the authors also attended a roundtable discussion at the annual Medical Library Association conference in 2015 to consider how other librarians were implementing mobile technology.

Initially, twenty-five apps were selected, which fit loosely into four categories: anatomy, diagnostic, education, and medication. The differential in cost ranged from free to $2,100 to purchase thirty instances of an app through Apple's Volume Purchase Program for Education. The most expensive app was the American Psychiatric Association's *Diagnostic and Statistical Manual of Mental Disorders* (DSM-V), which we ended up not choosing.

The selected anatomy apps included organ and musculoskeletal systems, thus serving the needs of the schools both of Nursing and of Health and Rehabilitation Sciences. These apps included Anatomy 4D, Heart Pro III, iPhysio, Muscle Premium, and Sobotta Anatomy Atlas. The diagnostic app category of free and purchased products included the Electronic Preventive Services Selector (ePSS), the Johns Hopkins ABX Guide, Kidometer, Lab Values, Medscape, the Merck Manual, Nurse-Tabs, and Nursing Essentials. This category also included the evidence-based medicine tools that were already available through our purchased databases: First Consult, DynaMed, UpToDate, and VisualDx. The education apps could be used for training, tutoring, and simulation of patient information. These were ECG Rhythm Tutor, MedLab Tutor, and PALS Advisor. Several of the anatomy apps could also be used in this way. The medication app category included Micromedex Drug Reference, Nurse's Drug Handbook, and Epocrates Essentials. Micromedex requires an annual subscription and was the only one of the three that was purchased.

## Instruction with iPads

Since the apps were available only on the iPads for students enrolled in one of the health sciences programs, students were introduced to these products through library instruction during the spring 2016 semester. This brief pilot period limited the number of instructional opportunities and collaborations with faculty. Consequently, the liaison librarian for the School of Health and Rehabilitation Sciences (SHRS) had an opportunity to use the iPads with a group of high school students. The liaison librarian for Informatics and Computing deployed the iPads in a library orientation training session. The liaison librarian for the School of Nursing introduced the iPads and health-related apps to graduate nursing students in one of her library instructional sessions.

### High School Students

The liaison for the School of Health and Rehabilitation Sciences (SHRS) participated in the Indiana University Health Careers Opportunity Program (IU-HCOP) Saturday Senior Academy in order to provide research skills instruction and library services for senior high school students and instructors. The students had five library instructional sessions during the program and were introduced to the iPads in their first library instruction session. The University Library allowed the IU-HCOP coordinator to borrow iPads so that all instructors could utilize iPads in their sessions.

SHRS initiated the IU-HCOP to help disadvantaged students in acquiring academic and social skills so that they can successfully complete their health professions programs. SHRS believes that these efforts can lead to increased diversity in the health professions (SHRS, n.d.). As part of IU-HCOP, the Saturday Senior Academy was first launched in January 2016 with fifteen senior high school students (at the end of the program there were nineteen students) from Indianapolis public schools or Marion County schools. These students came to the IUPUI campus every Saturday until June 2016. The program consisted of math, science, and writing classes, college prep workshops, health career workshops, and field trips. In addition, there were group projects on health-care research using IUPUI University Library resources.

As part of this high school session, a survey was conducted on the first day to identify students' current level of familiarity with an iPad. Almost half of the students responded that they had already used an iPad in their high school classroom, although only 30 percent of students indicated that they had their own iPad at home. The majority of students strongly believed that the use of an iPad would improve their learning in class. The students reported that they felt an iPad would not only help them learn, remember, and participate, but also that it would serve as a general learning aid. In addition, fourteen students stated that they would like to have an iPad to use as a tool in their classes every day (figure 11.1).

**FIGURE 11.1**
Students' perspective on iPad as a learning tool.

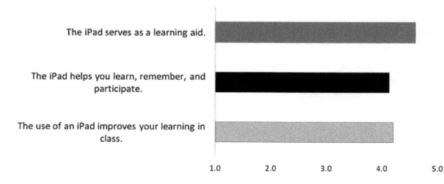

Of the five library instructional sessions, the high school students had one session dedicated to learning how to use an iPad and exploring health-care- or health-sciences-related apps available on the iPads. Since the high school students were not affiliated with IUPUI, they did not have IUPUI usernames and passphrases, so the liaison librarian created generic accounts for individual students before instruction.

It was obvious that the students did not have any difficulties operating an iPad. They already knew how to log in and how to open and close an app, but they knew very little, if anything, about educational or health-related apps. The students expressed their excitement about using these apps with the iPad, especially Anatomy 4D. The librarian demonstrated how to use Anatomy 4D first through a paper projector since a connector from the iPad to the computer was not available at the time. Students said that the app was helpful, interactive, and visually attractive and made it easy to understand the human body and heart models. However, since only three 4D images were available, the students wished that there were more 4D images to play with and learn from. The librarian also introduced other apps, such as Nursing Drug Handbook, MedLab Tutor, UpToDate, and DynaMed, to the students. Most of these apps required students to create their own username and password, to which the students were unreceptive. Although the librarian demonstrated how to create an account and how to answer medical questions using UpToDate, the students simply watched the librarian use the app and were not interested in participating, saying that that they did not want to create an individual account.

## *Undergraduate Students*

None of the librarians were able to use the iPads with a typical class of undergraduate students during the period of the pilot. In lieu of an undergraduate course, Willie Miller, liaison librarian for the School of Informatics and Computing, used the iPads during a library tour training session for thirty undergraduate orientation leaders. This training lasted thirty minutes, and the students used the iPads' native web browser, Safari, to navigate University Library's website. Miller used the website to identify key information about the library's services. The training was a combination of lecture and discussion, with iPads being used to help answer questions about the library.

As previously noted, Miller used iPads with undergraduate students during the 2010–2013 academic years. During that study, students appeared to be captivated by the devices and reported high levels of engagement; however, Miller found that a small number of students were distracted from lecture because of their fascination with the iPads. In these early classes, Miller and his colleagues would teach students to how the iPads worked, going over what the buttons did and how to access the materials needed for the activities. This process would take about twenty to sixty minutes, depending upon the instructor.

Miller noted that student behavior had changed since the iPads were initially released. None of the students were distracted from training by the iPads; further, the students did not need instruction on using the iPads as they already knew how they worked. Though this may not be the case for every undergraduate, the popularity of iOS reduces the learning curve for many students and other users.

Due to time limitations, Miller used informal assessment to learn more about the students' experience using iPads. Before the end of the training, he asked the students questions from the questionnaire Lee constructed. All of the students reported having used the device before. Most of the students agreed that an iPad helped them learn, remember, and participate. Several of the students expressed a desire to use an iPad during the classes. Some students asked if the iPads could be used to download e-books. In answering the question, Miller relayed the complexities of academic e-book publishing. The university library currently has twenty-one platforms from which e-books can be viewed or downloaded.

## *Graduate Students*

Caitlin Pike, liaison librarian for the School of Nursing, demonstrated the iPads in an elective, graduate-level Scholarly Projects class. Pike is embedded in this class, teaching two four-hour intensive workshop sessions per semester to assist students in developing research topics, learning database searching skills, and understanding what resources are available to them once they leave the university. As a result of this, the course seemed best suited to distributing the iPads for testing in a sandbox-type environment.

The librarian handed out the iPads on her second day of class, asking several informal questions as the students tried logging in. These included "Does anyone currently use tablets at your hospital?" and "If so, what apps have you used?"

The majority of students in the class were pediatric nursing specialists, so the apps demonstrated were chosen based on their potential usefulness to pediatric nurses. It was difficult to select which apps to demonstrate, as many of them required individual accounts. The librarians are still working with IT to determine the best course of action regarding individual accounts. We are considering either creating placeholder accounts for the students to use while on our iPads or having the students create their own individual accounts, which they can then transfer to their own devices.

The students worked with ePSS, PALS Advisor, ACLS Rhythm Tutor, Kidometer, and Sobotta Anatomy Atlas. They also had the opportunity to explore some of the other apps, including the Merck Manual, Heart Pro III, and Micromedex Drug Reference. According to the postdemonstration survey, 50 percent of the students said they would be very likely to use the ePSS app, the Merck Manual, Nursing Drug Handbook, and Micromedex Drug Reference. In addition, 66 percent of the students said they preferred to use the apps on a mobile device instead of a laptop. The students reported barriers to using the iPads in the classroom as having difficulty signing into the iPads and the subsequent need to have individual accounts for several apps. One student also expressed a need for more time during the demo to explore. Other barriers to using the iPads in the future included storage space, security, and cost.

## Barriers or Issues
### Technical Issues
After a student logs in using his or her IUPUI username and passphrase, he or she is able to download additional free apps like Dropbox or Google Drive via an app catalog maintained and approved by UITS. The app catalog, like the App Store, is a list of all of the apps available in our collection, which students can install on the shared iPads with their IUPUI username. The EMM suites and the Single App lock feature enabled us to facilitate this process as easily as possible. However, it took us almost six months to determine the best process before deploying the project in the classroom. This delay effectively shortened the pilot period to the spring 2016 semester.

There were also several problems with UITS AirWatch Mobile Device Management system, which caused the iPads to lock up after login. In addition, during one of the library instruction sessions, students could not log in to the iPads with their IUPUI username and passphrases, so the librarian had to use her username and passphrase to log in. As mentioned above, most of the apps require the user to create individual accounts in spite of institutional subscription. The authors are in the process of ascertaining if students would prefer the creation of generic accounts or if the creation of their own accounts would better enhance the app experience.

### Effective Teaching Methods
There are many sources of information available on the best apps for health sciences students. However, very few mention how to teach with them effectively or how to assess mobile literacy. Many students do not need to learn how to use apps since the apps are designed to be easy to use. However, they still need to learn how to apply information literacy concepts such as critical-thinking skills and effectively evaluate health information when they use health-related apps like UpToDate or DynaMed. Because it was the first time using iPads during a library instruction session, the two librarians had difficulty finding an effective teaching format and deciding what to teach (i.e., demonstrating how to use apps for research or just introducing useful apps). In addition, all three librarians tried to locate existing assessment tools related to teaching iPads and apps in library instruction work-

shops. However, they could not find one. Consequently, they had to develop their own survey assessment for this pilot project.

### Availability of Hardware

Currently, the purchased apps are available only on the shared iPads. Many academic libraries, such as the Indiana University Ruth Lilly Medical Library, McGill University Library, and the University of Toronto Libraries, have a web presence to list useful free apps or library subscription–based apps. The three librarians are also in the process of creating a guide or webpage to list available apps, but we would also like to add apps to the library catalog so that users may access them like e-books or articles through their mobile devices. However, there are many barriers to making this happen. First, students have to install the apps. Second, they may need to create an individual account before using an app. After such complicated steps, many students might give up on library subscription–based apps.

### Apps Purchasing Workflow

Determining responsibility for purchasing mobile apps was a challenge. Traditionally, the acquisitions team handles all of the collection acquisition processes. However, since apps are unique and different products, it was not clear how to proceed. At our institution, CST is responsible for not only maintaining mobile equipment, but also purchasing apps. Moreover, traditional collection development funds were not used for purchasing apps; instead, the authors have used the annual budget for technology. The authors believe that libraries have to think about best practices for app purchasing workflows as apps become increasingly popular.

### How to Maintain Apps as Collections

Due to the limitation of 16 GB storage, we had to decide which apps to add to the iPads and which apps to remove. Even though EMM suites provide apps usage analytics, it is a hard decision to make as we do not have enough data for this semester to form a complete picture.

## Conclusion

Mobile devices and apps create new possibilities by linking to alternative content, like audio, video, and images. They also help break down digital barriers since they are less expensive than computers. Hennig (2014) states, "Compared to the cost of desktop and laptop computers, mobile devices are generally less expensive" (p.

3). Furthermore, apps have become popular and ubiquitous. The editors of *MIT Technology Review* chose Mobile Collaboration as one of the Top Ten 2014 Breakthrough Technologies, which included the collaboration apps Box, CloudOn, Dropbox, Google Drive, Microsoft's OneDrive, and Quip (Greenwald, 2014). TED was the Webby Award's (2015) Mobile Sites and Apps winner in the category of Education and Reference. As Ally and Prieto-Blázquez state, "In the future, mobile devices will look completely different from today's; hence, higher education must plan to deliver education to meet the demands of new generations of students" (p. 144).

Along with new technologies, librarians have traditionally sought for better ways to connect their users with library resources and services. Since mobile devices and apps have become important tools for finding information, many libraries have adapted and integrated them into either their services or library instruction programs. Based on the pilot program, the authors discovered that it is important to not only teach students how to effectively use the apps in an educational setting, but also to make the apps easily accessible through both shared and personal iPads. The authors believe that apps should be considered a part of libraries' collections, and users should be introduced to them through a library's website or guides. This would allow libraries to connect users with apps across platforms and devices.

# References

Ally, M., & Prieto-Blázquez, J. (2014). What is the future of mobile learning in education? Mobile learning applications in higher education [special section]. *Revista De Universidad y Sociedad del Conocimiento, 11*(1), 142–151.

Barker, K., & Knapp, M. (2014). *Appsense makes the patrons grow fonder: Mobile resources in the health sciences* [Webcast].

Belardi, B. (2015). *Report: New McGraw-Hill Education research finds more than 80 percent of students use mobile technology to study.* Retrieved from http://www.prnewswire.com/news-releases/report-new-mcgraw-hill-education-research-finds-more-than-80-percent-of-students-use-mobile-technology-to-study-300047130.html.

Capdarest-Arest, N. A. (2013). Implementing a tablet circulation program on a shoestring. *Journal of the Medical Library Association: JMLA, 101*(3), 220–224. doi:10.3163/1536-5050.101.3.013.

Deloitte Center for Health Solutions. (2012). *mHealth in an mWorld: How mobile technology is transforming health care.* Retrieved from http://www2.deloitte.com/us/en/pages/life-sciences-and-health-care/articles/center-for-health-solutions-mhealth-in-an-mworld.html.

Duncan, V., Kumaran, M., Lê, M.-L., & Murphy, S. (2013). Mobile devices and their use in library professional practice: The health librarian and the ipad. *Journal of Electronic Resources Librarianship, 25*(3), 201–214. doi:10.1080/1941126X.2013.813304.

Gillum, S., & Chiplock, A. (2014). How to build successful iPad programs in health science libraries: A tale of two libraries. *Journal of Electronic Resources in Medical Libraries, 11*(1), 29–38.

Hennig, N. (2014). *Apps for Librarians: Using the best mobile technology to educate, create and engage.* Santa Barbara, California: Libraries Unlimited.

Lotts, M., & Graves, S. (2011). Using the iPad for reference services librarians go mobile. *College & Research Libraries News, 72*(4), 217–220.

Maloney, M. M., & Wells, V. A. (2012). iPads to enhance user engagement during reference interactions. *Library Technology Reports, 48*(8), 11–16.

Massis, B. E. (2013). From iPads to fishing rods: Checking out library materials. *New Library World, 114*(1/2), 80–83.

McCabe, K. M., & MacDonald, J. R. (2011). Roaming reference: Reinvigorating reference through point of need service. *Partnership: The Canadian Journal of Library and Information Practice and Research, 6*(2).

Poll, H. (2015). *Pearson student mobile device survey 2015.* [Presentation Slides]. Retrieved from http://www.pearsoned.com/wp-content/uploads/2015-Pearson-Student-Mobile-Device-Survey-College.pdf.

Payne, K. F. B., Wharrad, H., & Watts, K. (2012). Smartphone and medical related App use among medical students and junior doctors in the United Kingdom (UK): A regional survey. *Medical Informatics and Decision Making.* Retrieved from http://www.biomedcentral.com/1472-6947/12/121.

Rossing, J. P., Miller, W., Cecil, A. K., & Stamper, S. (2012). iLearning: The future of higher education? Student perceptions on learning with mobile tablets. *Journal for the Scholarship of Teaching and Learning, 12*(2), 1–26.

SHRS (School of Health and Rehabilitation Sciences). (n.d.) *Indiana University Health Careers Opportunity Program.* Retrieved from https://shrs.iupui.edu/about/IU-HCOP/index.html.

Shurtz, S., Halling, T. D., & McKay, B. (2011). Assessing user preferences to circulate iPads in an academic medical library. *Journal of Electronic Resources in Medical Libraries, 8*(4), 311–324.

Smith, A. (2015). *U.S. Smartphone use in 2015.* Retrieved from Pew Research Center: http://www.pewinternet.org/2015/04/01/us-smartphone-use-in-2015/.

Spence, E. (2016). Microsoft's smartphone strategy is simple: It will change the rules. *Forbes.* Retrieved from Microsoft's Smartphone Strategy Is Simple: It Will Change The Rules website: http://www.forbes.com/sites/ewanspence/2016/02/14/microsoft-windows-10-mobile-strategy/#205ced7e5d2f.

Tao, D., McCarthy, P. G., Krieger, M. M., & Webb, A. B. (2009). The mobile reference service: A case study of an onsite reference service program at the school of public health. *Journal of the Medical Library Association: JMLA, 97*(1), 34–40. doi:10.3163/1536-5050.97.1.006.

CHAPTER 12*

# Tablets on the Floor
## A Peer-to-Peer Roaming Service at Atkins Library

*Barry Falls, Beth Martin, and Abby Moore*

## Introduction

Circulation services at J. Murrey Atkins Library were once very desk-centric. The patrons came to the circulation desk for all kinds of help including printing, directions, circulation, and reference help. The Atkins Library is open twenty-four hours a day for five days of the week, starting Sunday at 11:00 a.m. and closing Friday at 8:00 p.m., and we are also open Saturdays from 10:00 a.m. to 8:00 p.m. The library is ten floors high, and many students do not want to leave their belongings (or pack them up) to ask a question at the circulation desk.

The Research and Instruction Services (RIS) desk already had chat and text messaging features. However, the desk was not open at all hours. It was not reasonable to staff the RIS desk during hours where there was clearly not a need for the service; therefore, it closes at 10:00 p.m. during the week. The circulation desk has at least two people scheduled on the desk at all times, but RIS normally has one team member scheduled per shift, and the questions he or she answers are not as transactional. In other words, reference librarians can expect to spend more time with a patron discussing in-depth research questions, while circulation questions rarely require that length of time.

Atkins had also recently remodeled the desk area and removed a large information desk to make way for more seating. This put even more demand on the

---

* This work is licensed under a Creative Commons Attribution-ShareAlike 4.0 License, CC BY-SA (https://creativecommons.org/licenses/by-sa/4.0/).

circulation desk to answer these in-depth consultations. It was time to consider ways to expand the general circulation and information services beyond a desk.

In 2015, an evening circulation team member mentioned, in passing, that they were using students' phones to help find information and that it would be helpful to have a tablet when they were walking around the library. It was a simple request, but it opened up a lot of possibilities for changing the service model. In addition, a study of service desk analytics was done, and the results indicated that many of the questions asked at both desks were often general in nature, such as how to find a book, directions, or printing issues. These questions didn't require a desk and could be answered by any available team member, including, potentially, a roaming team member. Roaming or roving services are not new to research and reference librarians, but there was not a model for this in circulation. Given the proven needs of our patrons, such services seemed to be a good fit for our community.

## Logistics: Making It Happen

The first step in the plan to expand circulation and research services was to purchase iPads so that staff could carry them around the floor when not at the desk. This step seemed reasonable, and with the support of the interim University Librarian, several devices were purchased. Either an iPad or an iPad Mini was ordered for every staff member who worked in circulation. However, the circulation team realized that carrying a tablet connected to the Wi-Fi around the library floor might not be sufficient. Quite often, when staff members were away from the desk, they had their other job duties to fulfill and were rarely just walking around the building. The head of Public Services realized that team members might need to be designated specifically to fill this new role. A student employee took over the roaming program and started to assist in coordinating a group of students to walk around the building and offer on-site help. Students would be trained to help their peers on the library floor, not behind the desk, in order to develop a peer-to-peer service model.

The RIS department offers a service called Ask Atkins, which includes chat, text, phone, and e-mail reference in addition to its traditional office consultations. Upon consultation with the Ask Atkins team, both the Ask Atkins staff and the library administration approved the roaming project. It was considered a great fit with this established program. This collaboration was also a way to bring the two largest public service groups together: Access Services and Research and Instruction Services. These departments now had greater lines of communication and a better understanding of each other's areas of expertise. An Ask Atkins logo was created by the in-house graphic designer as a way to brand the service on the website and on nametags.

Naturally, this project led to new questions such as "Can roamers look up patron accounts from an iPad?" and "Can roamers even check out books for patrons?" Atkins Library became a WorldShare Management Services library in July of 2013, which means that the integrated library system (ILS) is a cloud-based system. Because of the implementation of this system, tablets could be used not only to surf the Internet, but also to look up patron accounts and answer very specific questions. This mobile ability was an exciting development that opened up new possibilities, such as the deployment of wireless barcode scanners in order to check out items for patrons at their point of need.

## Pilot Project

A pilot project began in October 2014 and ran through January 2015, with one employee from Access Services roaming twice per weekday for one-hour roaming shifts. This roamer walked around each of the library's ten floors, carrying an iPad loaded with a survey created in Google Forms to document interactions with patrons.

The roaming process provided valuable opportunities to rethink the organization's service model in order to move beyond the image of the traditional academic library service desk. Previously, Atkins Library's contact points with patrons were either desk-based (reference desk and circulation desk) or virtual (e-mail, chat, and phone). Roaming services could be positioned to supplement the library's other face-to-face patron transactions by increasing staff visibility and decreasing the anxiety patrons sometimes associate with approaching librarians.

Since at least the 1970s, librarians have understood the ramifications of library staff being tied to desks. Research by Swope and Katzer (1972) emphasized the potential for roamers to address the needs of "non-askers," that is, patrons who are unwilling to seek out a librarian for assistance. The same study found that approximately 75 percent of patrons were unwilling to direct their questions to a librarian due to past experiences with librarians or because they didn't want to bother the librarian. Roaming services aim to address this patron group by meeting them where they are to address whatever questions they may have.

Historically, successful roaming services rely on the technology to support the service, as well as relying on the organization's full commitment to developing and modifying the service to fit the unique needs of the community. There is no one-size-fits-all approach to developing a roaming library service. Thus, the pilot project also aimed to assess how the roaming service might be positioned to meet specific needs.

Data was gathered on several components of each patron interaction. When and where did the interaction take place? Who initiated it? What type of question or inquiry was addressed? How long did it take to assist the patron? How would the roamer rate his or her helpfulness in each interaction? And how would the roamer rate the patron's interest in the service?

The results of the survey suggested that there was a demand for a point-of-need service. Nearly half of the patrons approached during the pilot needed assistance. Of questions, 30 percent were quick reference requests, 6 percent were related to the library printing service, and the remaining questions involved reshelving unwanted items or information about circulation policies. The most assistance was needed near library printers and study rooms.

Most patron interactions were very brief, lasting around five minutes or less. This discovery was a positive outcome as the Atkins Library was positioning the service to be quick and accessible. Any in-depth reference or research inquiries were directed to the research service desk. Of all interactions, 90 percent were completed without the need to refer patrons to another service desk. Nearly all patrons were open to assistance, even if they were initially apprehensive about being approached.

The roamers initiated 69 percent of all interactions, while the patrons initiated 31 percent. This was due, in part, to a lack of identification and public knowledge of the service. Lack of visible identification posed various issues during the pilot, as being approached by an unidentified library worker often seemed to make patrons uneasy. Consequently, library name tags were distributed to all roamers in time for the soft rollout in summer 2015. In keeping with the cross-departmental and collaborative nature of the service, Ask Atkins–branded name tags were distributed to all employees who assist patrons face-to-face, including those in the Access Services, Research and Instructional Services, and Collections Maintenance departments.

One important discovery during the pilot project was that the library's Wi-Fi infrastructure was not as robust as needed in order to ensure that Internet access was always available to roamers. There were frequent drops in connectivity, especially as the roamers moved through the upper floors of the library. This impeded on roamers' ability to assist with most patron inquiries. In response, data was collected on how frequently roamers lost connection to the building's Wi-Fi. This data provided the library dean with additional evidence to argue the case for an updated wireless system for the library building.

The pilot project also provided some insights into the differences between academic and social spaces in the library. While not deliberately designated as such, certain spaces in the library (mainly a corner of the third floor) were used primarily as social space for gatherings by various fraternities, sororities, and other student organizations. Students in these areas were generally less likely to require assistance from roamers. Similar discoveries from the pilot mirrored the research of Schmehl-Hines (2007), whose studies on outpost reference suggest that "the more academic a space was, the more likely it was to be a successful outpost location" (p. 13).

## *Roaming Technology and Software*

Technology played an integral role in the pilot and required extensive research, with particular attention given to how technological advancements have influenced innovations in past library roaming services. Library professionals began developing roving reference services in the pre-mobile era. In fall 1989, Boston College librarians began experimenting with their own roving service in order to address the needs posed by an "explosion of electronic resources" (Bregman, 1992, p. 634); staff would leave the service desk to assist patrons with using Boston College's new virtual research tools.

Since that time, roving reference staff at various institutions began using laptops on carts, which allowed them to search the library catalog and the library directory on the spot, and even check out books at the point of need. However, this provided some mobility challenges. "Today, tablets are the preferred instruments for many library professionals who implement their own roaming services to conduct their own seamless reference interactions," (Hibner, 2005). With countless brands and models of tablets available, the pilot also attempted to experiment with different tablets in order to determine which would best fit the Atkins Library roaming service.

The roamer for the pilot project experimented with three different tablets: a fourth-generation iPad, an iPad Mini 3, and a second-generation Amazon Fire. The Amazon Fire tablet, which was marketed as a strong competitor to the iPad (Letzing, 2011), contained many of the key functions roamers would require. However, it was also very slow and unintuitive, and connecting to the library Wi-Fi was frequently difficult. The iPad Mini 3 was a technical step up from the Amazon Fire, but the small screen (and proportionately small keypad) made searching for books in our catalog less than ideal for a service that was designed to be quick and effortless. In the end, the pilot project demonstrated that the fourth-generation iPad, which has a 9.7-inch (diagonal) screen and weighed 1.44 pounds, was ultimately the roamer's tablet of choice.

One issue that surfaced during the pilot project was that circulation staff needed a way to get in touch with the roamers from the desk when the roamers were on the floor. The head of Public Services met with the RIS team, and subsequently circulation was integrated into the chat system, LibraryH3lp (https://libraryh3lp.com) LibraryH3lp allowed us to create a separate virtual chat room for our circulation team, which at the time was spread throughout the building. Previously, anyone at the circulation desk had to call random numbers until help was found, but with the integration of the chat function, all of circulation, including our branch library, interlibrary loan, and Collections Maintenance, was added to the chat room. A patron could come to the desk or stop a roamer and get immediate answers to questions even if the questions were not circulation questions. The service was expanded so that members of the RIS team who were monitoring chat or working with a patron could get on the system and ask a quick question of circulation staff. By expanding the tool beyond RIS, we improved services and staff met students at their point of need. The roaming service was created to meet student needs in a variety of spaces and to expand services beyond a traditional circulation desk.

## Hiring and Training

The data collected from the pilot project helped to inform the steps for transitioning from pilot to launch. A soft rollout occurred in the summer of 2015, with the official launch scheduled for the beginning of the fall semester. Before the launch, the library needed to hire and train new student roamers. Much of the research conducted on similar point-of-need services indicated that experienced reference librarians often staffed these services. However, the Atkins Library roaming service differed in that it was staffed exclusively by student workers.

Employing only student workers did present some disadvantages. For one, student employees require more in-depth reference training and are less familiar with the overall operations of the library system. Additionally, their course load can pose scheduling challenges. Shifting extracurricular activities meant adjusting schedules on a week-by-week basis, and during the times when most students are in class, finding a roamer to fill shifts could be difficult, and sometimes even impossible.

Despite these challenges, the benefits of staffing the roaming service with students were nonetheless numerous and compelling. Student staff "understand the demands placed on the students and their information needs on a more personal and empathetic level than the professional or technical library staff" (Neal, Ajamie, Harmon, Kellerby, & Schweikhard, 2010). Student workers may also be more comfortable with the technology used for the service. In addition, student workers can help to alleviate the library anxiety that some students experience when approaching the library professionals or permanent staff (Jiao, Q. G., & Onwuegbuzie, A. J., 1997; Mellon, 1986).

Hiring the right student workers for the rollout of the roaming service required careful planning. While supervisors oversee student workers at the desk, roamers work independently and with minimal supervision. During these hour-long walks, roamers must be proactive in their search for patron inquiries. Thus, the Roaming Services Coordinator carefully searched for student workers who were highly service-oriented and self-reliant.

## Training Development

Using the ADDIE (Analysis, Design, Develop, Implement, Evaluate) instructional design method, the library had already created a training plan to train reference desk workers on the five steps of the reference interview. This workshop was adapted into a training workshop for the peer-to-peer service program. It is important to note that roamers were trained at the reference desk and also at the circulation desk. Since there are two facets of reference training, content and behavior (Moysa, 2004), much of the training that was already included in the reference training workshop would apply to all interactions on the floor, because no matter what kind of question the roamers are asked, their behavior must remain consistent. According to Ward (2003), "Training staff properly for reference work involves learning not only lists of resources, but also appropriate behaviors that will ensure they understand and meet users' needs" (p. 46).

## TRAINING SESSIONS

The reference training plan included three types of training. The roamers met with the Education Librarian, who designed the training, for an introduction to reference services. Roamers then sat individually with a reference librarian at the reference desk for a practical look at what happens during a typical reference desk shift. Ideally, the roamers would be introduced to the reference interview during these training sessions. Finally, all of the roamers participated in a reference-training workshop, which was facilitated by the Education Librarian and the Head of Reference Services.

## FACE-TO-FACE TRAINING

According to the Association of Research Libraries (ARL), a reference transaction is "an information contact that involves the knowledge, use, recommendations, interpretation, or instruction in the use of one or more information sources by a member of the library staff" (2008), p. ). As an introduction to what reference service is all about, roamers were given this definition. The subsequent discussion about reference services led to an overview of what happens at the reference desk:

- Teach patrons how to use library resources.
- Teach patrons how to locate what they are looking for.
- Find answers to patrons' questions using library resources and other resources.
- Refer patrons to liaison librarians.
- Explain library policies and procedures.
- Offer reference service in person or via phone, e-mail, text, or chat.

The purpose of this reference training was not to train the roamers to work at the reference desk, but simply to introduce them to reference services and to prepare them for basic reference interactions. Before the training was designed, special consideration was given to what reference skills roamers would need most. The list included

- Locating books in the catalog and on the shelf
- Locating articles using databases
- Referring students to liaison librarians

At the end of the meeting, the roamers were introduced to the concept of the reference interview.

### ON-THE-DESK TRAINING

There is no better place to see the reference interview in action than at the reference desk. Each roamer-in-training sat with a librarian at the reference desk for at least one hour. Desk training was scheduled during peak reference hours. However, since this training occurred during the summer months, peak hours were somewhat slow. Most of the reference interactions the trainees observed were not real-world reference questions as more often than not, patrons came to the desk simply to ask printing and directional questions. Since the trainees would likely have to answer these types of questions anyway, their time spent at the desk was still beneficial. During the training workshop, roamers were provided with additional basic customer service and reference training.

## Implementation

The rollout of the roaming service occurred during the fall 2015 semester. It launched with minimal marketing initially, with five undergraduate student workers who spent half of their shift working at the circulation desk and the other half roaming. Roamers worked from 9:00 a.m. to 6:00 p.m. Monday through Friday. They roamed from the ground floor to the eighth floor without any standardized route. Though the project had moved out of the pilot period and into an official rollout, the same approach to experimentation remained. While roamers were more publically available to library patrons, the service's potential was far from fully realized. Roamers were encouraged to take initiative and to report possible innovations for the service.

Roamers did not use any special iPad applications. The web browser, Safari, was the only preinstalled application. Four Safari tabs were always open during roaming sessions: the library homepage (which also contains a search bar to the library catalog), an analytics page (for logging statistics on each patron interaction), LibraryH3lp (our web chat service), and WorldShare Management Services (our resource circulation program). The iPads also have other sites bookmarked, such as the library directory, maps of library floors, study room booking pages, and circulation policies.

Atkins Library had used LibraryH3lp for instant messaging chat for less than a year before the roaming pilot, but an account was set up for roamers in time for the fall launch. The staff in Circulation, Research and Instructional Services and

Collections Maintenance now use LibraryH3lp at Atkins Library to communicate with each other about quick inquiries, updates, and reminders related to public services. LibraryH3lp can also be used on iPads. This allows circulation desk staff to use the in-house chat feature to connect patrons who come to the desk for assistance with a roamer who can meet them in the stacks in order to provide additional assistance.

## The Student Library Advisory Board's Opinion on Roamers

In January 2016, the library worked with our student government association to create the Student Library Advisory Board or SLAB. A large part of the advisory board's mandate is to share information between the library and our students, as well as to elicit student feedback on library projects. The first order of business was to discuss the roaming service and the board's opinions on the project. Some of the students were not yet aware of the service but were very excited about the possibility of using it, while others indicated that they already had experience using the service. SLAB contributed marketing suggestions, such as the production of T-shirts for the roamers and the production of more flyers or posters to better advertise the new service. SLAB also provided valuable tips on where to tweet and e-mail information about the service in order to reach a significant number of students. This important information was used revitalize the marketing plan for the roaming service.

## Focus Groups

In the spring of 2016, the Atkins Library Usability Coordinator assessed the roaming service. The Usability Coordinator collaborated with the Roaming Services Coordinator, the Head of Assessment, and the Education Librarian to complete this assessment. After the first semester of the service, a survey was distributed to the roamers to assess their experiences. After the survey was completed, two student focus groups were planned to assess the service from the perspective of the user. Ten questions were asked during the focus groups. The findings revealed that many of our students had never used the roaming service because they were unaware of it and because they couldn't identify the roamers. The study resulted in four recommendations:
- Increase user awareness.
- Increase user access to roamers.
- Evaluate the growth of the roaming service.
- Develop ongoing training for the roamers.

These recommendations were all taken into account for the new fall 2016 service model, and the team will perform ongoing assessment to improve the service.

## Marketing

Marketing this program could be a key component of success. However, the roaming team did not include a marketing person in the discussions until the service was already underway. Based on comments from both the roamers and the Student Library Advisory Board, bright T-shirts that would make roamers easy to identify in the stacks were considered an important part of the marketing campaign. The project team worked with the library marketing and design staff to create T-shirts that displayed the project logo and were easy to see across the library. Additionally, the communications director wrote a story for the library website to describe the service, and the graphic designer created posters for display around the library. In the future, the marketing team will be included in the preliminary phases of all of our projects to facilitate improved dissemination of information to the students.

## Hours of Operation

The initial rollout placed the roamers on the floor from 8:00 a.m. until 4:00 p.m.; however, this was not a data-driven decision, but one based on instinct and a change from the original scheduled proposed. Upon subsequently looking at the circulation and reference service statistics, the team realized that the ideal times in the library for this type of service would be from 10:00 a.m. or 11:00 a.m. until 8:00 p.m. This meant that roamers on duty in the early hours would have few questions to answer and there would be students who needed services when roamers were not available. Based on this data analysis, the service's hours were changed in the second semester. The roamers worked from noon to 8:00 p.m. Monday through Thursday and 10:00 a.m. to 6:00 p.m. on Friday, with future data analysis planned to investigate the possibility of adopting weekend hours for the service or any appropriate changes that need to be made to the service's initial operating hours. Figure 12.1 shows the percentage of types of questions roamers received by the hour over the spring 2016 semester. It became apparent that the students needed different types of help during different parts of the semester. For example, in the first two weeks of school, roamers dealt primarily with printing problems, while roamers working in weeks three and four saw more directional questions. In addition, reference questions begin to increase after week two and then to steadily increase throughout the semester. Exam time brings a renewed focus on printing problems as well as questions about group study rooms. Due to these insights, it was decided to have more students scheduled during the first week of classes and to ask regular staff to roam the floor to help with questions (particularly printing) during this time. This schedule will be scaled back later in the semester, and roamers will be assigned alternate duties as customer service demand dictates. For example, roamers can help to record student behaviors and space use for the assessment department when questions are less frequent since they are already on the floor and are able to observe students in a variety of library spaces. The patron assistance roaming duties always come first, but it is feasible to have roamers help with other work as needed.

**FIGURE 12.1**
Types of questions roamers were asked by the hour, spring semester 2016.

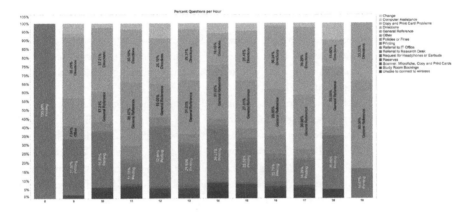

## Unexpected Benefits

An interesting outcome of the pilot was the larger impact that this project ended up having on the library in regard to technology upgrades. Many of the services that the roamers provide depend on being connected to the library's Wi-Fi. This connectivity proved troublesome in various areas of the building. These Wi-Fi "dead zones" made reference questions impossible to address. Roamers would often be forced to locate a nearby public computer in order to search the library catalog when they were unable to quickly reconnect to the Wi-Fi, which defeated the purpose of the service. Also, staying in touch with other public service staff (via LibraryH3lp) was equally problematic.

In September 2015, the roamers were tasked with documenting how frequently they lost connection to the library Wi-Fi while on their roaming shifts. The dean of the library used the resulting data to make the case to university administration and to the university IT department that Atkins Library was overdue for an updated Wi-Fi infrastructure. The dean's subsequent case to the administration was successful, and a new wireless access point upgrade was completed during the fall of 2015. A second assessment of the library's Wi-Fi network showed that that the updates resulted in roamer Wi-Fi drops being few and far between.

## Reflection and Next Steps

Informed by recommendations from the Usability Coordinator, the next steps for the program will aim to enhance the engagement, effectiveness, user awareness, and desirability of the roaming service in an effort to improve patron satisfaction and positively affect student academic success. These next steps are threefold:

1. Launch a vigorous marketing campaign that promotes the roaming service as accessible and useful.
2. Develop standardized instruction modules that improve the delivery of the service as well as keeping the roamers up-to-date on changes to library resources, services, and policies.
3. Continue to evaluate the service using a variety of assessment methodologies, including patron interaction statistics, employee surveys, and client focus groups.

The results of the client focus groups held to date confirmed that the roaming service's sparse marketing efforts hindered the service's overall reach. User awareness is key. The recommendations from the Usability Coordinator stressed the need for signage in the library promoting the roaming service. Fortunately, a newly redesigned library website, as well as new digital signage recently installed in high-traffic areas of the library, presented the ideal platforms for advertising the service. Furthermore, the start of the fall semester provides opportunities for new ways to market the roaming service, especially to new students—both transfer students and freshmen. Various events are held on campus to introduce incoming students to the university's programs, organizations, and resources. Plans are underway to promote the roaming services at freshman orientation events as well as inside the library with games, activities, and refreshments during the first week of the semester.

The training for the roamers was sufficient for the launch of the new program but will require extensive updates as the service matures. A complete training program with formative assessment activities and additional workshops must be implemented. In fact, an ongoing training program should be designed in which roamers, new Public Services hires, and student and temporary workers can participate. To supplement training modules and workshops, a training manual will be created that will include must-know information regarding Research Services and Circulation policies and procedures, along with information specific to the roaming program. Additionally, avenues must be created for communication between roamers and the trainers. For a training program to be successful, open avenues of conversation must be established, because, according to Drewitz (2013), new roamers "will be more comfortable asking questions" (p. 23), and training must be consistent and ongoing. Roamer training will soon evolve into something more like an ongoing training program and less like a one-shot instruction session.

Standardized training is also made necessary by new technological resources and services that the library now offers. Atkins Library is rolling out a Technology and Maker Space program, and part of this program will involve bringing roamers into this service area. Some of the new hiring will focus on recruiting students who have some technology knowledge in areas that the Atkins Library wants to support, including the Adobe Cloud Suite, Raspberry Pi, Arduino, Swivl, and other new technologies. Additionally, this new "technology help group" will work with roamers to learn about the mobile technology that the roamers use on the floor, as well as to support traditional library technology, such as microfilm readers. The group will also help with tools students use on a regular basis, such as the Microsoft Office suite and the Google Apps for Education products. While there is an explicit information technology service area within the library, it is staffed by the campus IT department, which helps students only with campus software and hardware, such as wireless access and password recovery. The new technology roaming help service, in conjunction with campus IT, will provide a complete service package to students—when and where they require assistance.

Finally, the long-term expansion of the roaming service relies on assessment strategies that must be implemented on a continual basis. When examined together, user statistics, employee surveys, and client focus groups provide a well-rounded perspective of a service. User statistics measure the overall effectiveness of the service, while the employee surveys and client focus groups provide insights on the employee and patron perceptions, respectively.

The goal of expanding general circulation and reference services beyond a service desk in the Atkins Library at UNC Charlotte has been met. However, changes and improvements to the service will be implemented in the coming months. Relying on data collected from both patrons and roamers and open communication between roamers and the Roaming Services Coordinator, the authors intend to explore patterns, inconsistencies, and other insights to plan and implement changes and improvements to the roaming service to ensure that the service, and the roamers themselves, will have a substantial impact on patron satisfaction and academic success.

# References

American Library Association. (2008). Definitions of reference: Reference and user services association. Retrieved from: http://www.ala.org/rusa/guidelines/definitionsreference.

Bregman, A. A., & Mento, B. (1992, November). Reference roving at Boston College: Point of use assistance to electronic resource users reduces stress. *College & Research Libraries*

*News, 53*(10), 634–637.

Drewitz, J. M. (2013). Training student workers: A win-win. *AALL Spectrum, 18*(2), 22–24.

Hibner, H. (n.d.). The Wireless Librarian: using tablet PCs for ultimate reference and customer service: a case study. *Library Hi Tech News, 22*(5), 19–22.

Jiao, Q. G., & Onwuegbuzie, A. J. (1997). Antecedents of library anxiety. *Library Quarterly, 67*(4), 372.

Mellon, C. A. (1986). Library anxiety: A grounded theory and its development. *College and Research Libraries, 47*(2), 160–165.

Moysa, S. (2004). Evaluation of customer service behaviour at the reference desk in an academic library. *Feliciter, 50*(2), 60–63.

Neal, R. E., Ajamie, L. F., Harmon, K. D., Kellerby, C. D., & Schweikhard, A. J. (2010). Peer education in the commons: A new approach to reference services. *Medical Reference Services Quarterly, 29*(4), 405–413.

Schmehl-Hines, S. (2007). Outpost reference: Meeting patrons on their own ground. *Pnla Quarterly, 72*(1), 12–13.

Swope, M.J., & Katzer, J. (1972). The silent majority: Why don't they ask questions? American Library Association, 161–166.

Ward, D. (2003). Using virtual reference transcripts for staff training. *Reference Services Review, 31*(1), 46–56.

CHAPTER 13*

# Using Proximity Beacons and the Physical Web to Promote Library Research and Instructional Services

*Jordan M. Nielsen and Keven M. Jeffery*

## Introduction

Modern academic libraries offer a wide array of resources that support the evolving curricular and research needs of their campus communities. In an age when library users have access to so much information, libraries must strike a balance between promoting independent inquiry and supporting users at their point of need. In the last two decades, this has prompted academic libraries to adopt services such as e-mail or chat reference, virtual office hours via web conferencing, online tutorials, and many other alternatives to traditional face-to-face reference and research support. Additionally, academic libraries have been expanding to include spaces such as writing centers, peer-to-peer tutoring, and makerspaces. Is it possible for libraries to provide support for these new spaces in alternative ways, similar to what they have done with other services? One solution may be to provide support through the combination of mobile devices and proximity beacons.

Proximity beacons transmit information using Bluetooth technology that can be picked up by mobile devices within a small area. This technology has

---

\* This work is licensed under a Creative Commons Attribution-NonCommercial-ShareAlike 4.0 License, CC BY-NC-SA (https://creativecommons.org/licenses/by-nc-sa/4.0/).

largely been used by brick-and-mortar businesses to push advertisements to customers based on their physical location in their stores. The leveraging of beacons to connect objects and spaces as an Internet of Things (IoT), also known as the Physical Web, is being promoted by leading companies like Google and Apple. Users are able to connect to supplemental information about an object using an app, or a physical web browser such as Google's Chrome browser. This allows organizations to contextualize a user's digital interactions in proximity to a particular physical space without requiring much effort from the user. These technologies represent a great opportunity for academic libraries seeking new ways to teach users about their spaces, services, and resources.

In the 2015–2016 academic year, librarians at San Diego State University (SDSU) implemented Eddystone proximity beacons and Physical Web technology in the University Library. Once librarians at SDSU made the decision to use these technologies, they then had to decide which areas in the library would house the beacons. The featured areas included the main entrance to the library, as well as the library's makerspace, financial data lab, writing center, and math tutoring center. For each featured area, a webpage was developed and its URL assigned to the corresponding proximity beacon. The webpages feature dynamic content about each of the featured spaces, including relevant images, videos and tutorials, news feeds, and portals to services such as chat reference and consultation scheduling. Once the areas were selected and the webpages were developed, the Eddystone beacons were deployed throughout the library. After the beacons were deployed, a promotional campaign began in the library to raise awareness of these technologies. This promotional campaign included digital and print signage throughout the library to make users aware of the new Physical Web links that could be explored, as well as banners on the library's website highlighting the new service.

This chapter will cover five distinct aspects of the proximity beacons project: background research, implementation, promotion, lessons learned, and next steps. The history of both proximity beacons and the Physical Web technology will be discussed, including their uses across various types of organizations. The selection of specific technologies and their implementation at SDSU will also be described in detail. As this technology is relatively new, it requires promotion in order to ensure user engagement. This chapter will, therefore, also describe strategies for making users aware of proximity beacons and Physical Web technology. There is also a learning curve when deploying any new technology, and lessons learned about the beacons, including technical hurdles, will be described. Finally, this chapter will conclude with suggestions for how beacons can be used in an academic library setting, including specific projects that are already taking shape at SDSU.

## Background
### *The Physical Web: The Internet of Things and Spaces*

The Internet of Things or Physical Web is taking off. From QR codes on sneakers leading to online running tips, to Internet-connected household thermostats, more and more objects, devices, and appliances have a presence on the Internet. Many researchers see this Physical Web eventually expanding to "enable an internet presence for any person, place or thing on the planet" (Want, Schilit, & Jenson, 2015, p. 29), expanding the Internet by as much as 100 fold, to billions of connected devices by the end of the decade (Jenson, Want, & Schilit, 2015). The SDSU Library had experimented with connecting its spaces to the Physical Web in the past. Starting in the spring 2014 semester, a self-guided mobile tour was created for students in a freshman general studies course, introducing them to library and university services. Posters were placed in eight areas of the library featuring QR codes, NFC (near-field communication) tags, and short URLs linking to an informational webpage about each area. This self-guided tour has been completed by over a thousand freshman students since its implementation, fully replacing the guided tours offered in previous years. This experience suggested that students would use tagging technologies to access supplemental information about spaces, as QR codes and NFC tag use were clearly the favorite access methods for this earlier project (Jeffery, Jarocki, & Baber, 2015).

Seeing the promise in leveraging the Physical Web to educate students, the library began to consider how it could "geofence" its spaces, a term that is commonly used when referring to location-based services and technology. Geofencing is the creation of a virtual area that, when entered or exited, triggers the "pushing" of a notification to a mobile device. The notifications that are pushed to mobile devices can feature information that is significant to the immediate location, thus providing appropriate contextual information to an individual (Greenwald, Hampel, Phadke, & Poosala, 2011). When exploring how geofencing could be implemented at SDSU, the library looked to proximity beacons. Proximity beacons are made possible by the Bluetooth 4.0 specification, using the Bluetooth Low Energy (BLE) subsystem. BLE broadcasts information on the same radio band as Wi-Fi, but in short spurts to conserve battery life (Faragher & Harle, 2015, p. 2418). These devices are small and unobtrusive, and they can be easily placed in designated geofenced areas.

Many businesses have begun experimenting with the use of beacons, especially retailers looking to engage customers shopping in their physical stores. In the summer of 2015, retail giant Target began testing proximity beacons in fifty of its stores (Monlloss, 2016). Lord and Taylor and Urban Outfitters use proximity beacons to connect with their mobile apps to offer in-store discounts to customers, and the Marriott hotel chain uses proximity beacons to identify guests and enhance the services that they are offered (Higginbotham, 2015). Beacons are also being used in major league sports to engage fans. For instance, Target Field, home to Major League Baseball's Minnesota Twins, has implemented beacon technology in their stadium to push content during sporting events (Brown, 2014).

There are a number of BLE beacon standards; however, for this project two standards from major technology companies were considered: Apple's iBeacon and Google's Eddystone-URL. iBeacon was announced by Apple in 2013 (Gotttipati, 2013), and Eddystone was launched by Google in 2014 (Garun, 2014). Apple's iBeacons broadcast a set of codes, a sixteen-digit UUID (universally unique identifier), a four-digit major code, and a four-digit minor code over BLE. These codes can be identified by apps specifically programmed to recognize the UUID codes and then serve up content designed for the beacon (BKON Connect, Inc., 2016). Any Bluetooth-capable device can make use of these codes. However, this requires an app designed to connect the beacon code to content. The information broadcast by iBeacons is therefore meaningful only to its related app (Jenson, Want, & Schilit, 2015).

Eddystone beacons, on the other hand, can transmit a UUID or a URL (uniform resource locator). The UUID transmitted by the Eddystone beacon partners with a specific app, like iBeacon. However, when Eddystone is set up to transmit a URL, this information can be seen and used by any app designed to detect the information being broadcast. There are a number of Physical Web browsers designed to detect these Eddystone URLs and display a notification to the user, allowing access to the content at that website address (BKON Connect, Inc., 2016). There are stand-alone Physical Web browser apps available for both Android and iOS. Google's Chrome browser itself can also be set up to detect Physical Web beacons on both platforms, as long as the destination URL of the beacon is using the secure HTTPS protocol (Phy.net, 2016).

## Implementation

For this project, Eddystone beacons were chosen. Initially, the library planned to include support for the iBeacon standard in Android and iOS apps designed by students from the university's master of computer science program as part of their thesis work; however, technical difficulties and time constraints removed this as an option. At the same time as the iBeacon implementation was proving problematic, Google released the Eddystone standard. The use of Eddystone beacons was attractive as they did not require the library to build an app given that any of the Physical Web–browsing apps supporting Eddystone beacons could be used.

The library considered promoting Google Chrome for the pilot, but this required the user to configure the browser to fully enable the technology in iOS. Instead, the library thought that promoting a single Physical Web browsing app for download could be more easily marketed. The library chose the Physical Web Browser by the Physical Web Team in part because this app used an easily identifiable Physical Web logo, which could be used in advertising and then readily identified by students in their app store. There are a number of providers of proximity beacon hardware, but the brand of Physical Web beacons chosen for this project was BKON. BKON devices were chosen as they required no ongoing contract and no associated app; were inexpensive, durable, and easy to manage; and made use of the Eddystone standard.

These BKON beacons broadcast a short URL on the PHY.net domain, which can then be associated with a destination URL using the free management interface on PHY.net, an associated website. The BKON devices require no direct programming, or even Internet access. The short URL is broadcast over Bluetooth, with the visitor's mobile device picking up this short URL and then connecting to the PHY.net website to resolve the destination website. By developing the informational landing pages in-house and selecting an à la carte beacon solution, this five beacon pilot project could be implemented for a little over $100 USD.

## *The Pilot Project*

To serve as a pilot project, five areas of the library were chosen to be featured using the proximity beacon technology. Three of the five areas chosen for this project are services within the library that are managed by external departments, including the writing center, math center, and financial data lab. The library's makerspace, Build IT, and the main entrance to the library were also chosen for this pilot project. These areas were chosen because they are student-focused spaces that are not currently featured on the library's mobile tour. The authors felt that this would cut down on any confusion with regard to the mobile tour while simultaneously promoting important spaces in the library. Along with testing proximity beacons and geofencing within the library, a primary goal for this pilot was to introduce students to areas within the library of which they might not be aware. By encouraging students to visit these five areas, the authors were hoping to increase the utilization of these spaces and services.

### LANDING PAGES

Content for the beacon websites was chosen based on the featured area. While each beacon landing page contained unique content, standardized elements were also included. The standardized elements included a video describing the featured area and a survey to gather feedback about the beacons and the relevance of the content on the landing page. The webpages were designed with PHP, and the first visit to any landing page set a thirty-day tracking cookie with an anonymous ID. This ID was used to identify a visitor if he or she visited another beacon and to log a return visit to a beacon during the same thirty days. Visitors also had the option of completing a short survey to give feedback on the beacons and the information on the supporting websites. The library was interested in how many students might visit the beacons in all five areas during the thirty-day trial. Statistics were also collected with the inclusion of a Google Analytics JavaScript on each page. Google Analytics is an easy way to obtain information on website visitors, such as device type, browser version, and operating system. A primary feature of each beacon landing page was a YouTube video instructing visitors about the geofenced areas in which they found themselves. These videos were scripted by a librarian, but featured the library's student liaison presenting the information. The library student liaison is a grant-funded position currently held by a sophomore student in the public relations program. The scripts ensured that the educational content was accurate, while a student delivering the information was thought to be a more welcoming method for the target student population than a librarian presenting the same information. The student liaison has created a number of videos for the library and has been given the affectionate moniker "Library Girl" by some of her fellow students who have seen the videos.

## MAIN LIBRARY ENTRANCE

SDSU's library is large and consists of two buildings that are connected via a lower level hallway. While there are two entrances, the main entrance is under a glass dome in the center of campus. The dome entrance represented a great opportunity to place a beacon and create a website for general library content. The beacon in the main entrance to the library connects to a website that features dynamic content from the library's website, including upcoming events, featured staff, services, resources, and social media updates. The website for this beacon also featured videos created by members of the library's Communication Team that changed each week and focused on a different theme connected to the library. This page also included the phone number for the library's Research Desk, with an invitation to students to send an SMS message for assistance.

## WRITING CENTER

The Writing Center is located in the library, but it is overseen by the Department of Rhetoric and Writing Studies. The center is run mostly by graduate students and serves as a place where students from any discipline can go to get help with their writing. For the Writing Center, the beacon website features multiple types of content. A brief video about the Writing Center was featured at the top of the beacon website, and this video offered an introduction to the Writing Center, including information about the purpose of the center, the hours of operation, and its contact information. The beacon website also featured information about booking a consultation (online or face-to-face) and a link to the online form used to schedule a consultation.

## WELLS FARGO FINANCIAL MARKETS LAB

The Wells Fargo Financial Markets Lab (WFFML) is a space within the library that contains twelve Bloomberg Professional terminals and a stock ticker. The WFFML is regularly used by SDSU's Department of Finance to conduct stock-trading courses. While the WFFML is heavily visited by students and faculty in business disciplines, it is open to all library visitors to encourage interdisciplinary use. The beacon website created for the WFFML contains a video providing an overview of the space and its features, a calendar with the hours of operation, links to the Department of Finance's website, and contact information for the lab managers. Another key feature of the WFFML's beacon is the links to Bloomberg Professional tutorials. Bloomberg Professional terminals are complex machines that have a steep learning curve. In order to assist users with learning the ins and outs of the machines, Bloomberg has created a certification program known as Bloomberg Essentials. One of the links on the WFFML's beacon page connects directly to a page describing the process for completing the Bloomberg Essentials certification.

## MATH AND STATISTICS LEARNING CENTER

The Math and Statistics Learning Center, managed by the Department of Mathematics, provides tutoring for students enrolled in math and statistics courses. The beacon website for the Math and Statistics Learning Center features each tutor's hours of availability, separated by two different calendars—one for statistics tutoring and one for math tutoring. The website also features links to the math-related research guides on the library's website and contact information for the Math Librarian.

## BUILD IT

Build IT is the library's makerspace and provides opportunities for anyone at SDSU to engage with 3-D printing and scanning, laser cutting, robotics, and virtual reality technologies. The space is primarily run by student volunteers known as Master Builders under the direction of the Engineering Librarian. The Master Builders go through extensive training so that they are able to safely operate all of the equipment in the space. Once they have completed this training, they are eligible to train other students on the machines. While the space is heavily used by engineering students, increasing numbers of students and faculty in other disciplines have been exploring the technologies Build IT has available. Build IT's beacon website features content from its social media accounts (Twitter and Instagram), the hours of operation, and information about becoming a Master Builder. Also, there is information about how the 3-D printing process works, including links to the Build IT website where 3-D printing jobs can be submitted.

## Promotion

Internet of Things technology is just beginning to take off. As a result, proximity beacons are relatively new and unknown. In order to raise awareness of the beacons and websites, multiple promotional avenues were explored. First, posters were created and placed in each of the areas containing a beacon. The posters contained information about the beacon, including information about how to download and use the Physical Web browser to access beacon content. Digital signs were also created using the same design scheme used in the posters. These digital signs are located throughout the library, including the heavily trafficked 24/7, Research Services/Circulation, and computer lab areas. The library's website, which has approximately 150,000 visitor sessions each month, also featured information about the beacons. A video was specifically created for the library's homepage that introduced the beacons and the Physical Web browser. The video also featured each of the areas that contained a beacon and encouraged students to explore all five areas.

A map was created for this pilot project by leveraging already-existing maps of the library. The map featured an overview of the library and specifically of the floors containing beacons. The beacon locations were denoted on the map by the presence of the Physical Web icon. The maps were printed and placed at each of the service points throughout the two buildings of the library, including the Research Services, Circulation, Print Services, and IT Help desks. The authors worked with supervisors at each desk to describe the beacon pilot project in case they received questions from students exploring the areas.

## Lessons Learned

University students have shown an unwillingness to use proximity tags like QR codes in the past. A 2011 study at the height of the QR Code marketing craze by the youth marketing firm Archrival found that 75 percent of students asked were "not likely" to scan a QR code, and only 21 percent could successfully scan a code when presented with one (Aguirre, D., Johnston, B. & Kohn, L., 2011). There will likely be similar problems with any technology that requires a specialized app, such as a Physical Web browser, but with Google committing to beacon support in its popular Chrome browser, the challenge with proximity beacons may become one of marketing, rather than technological adoption. It seems, looking at the number of pages prefetched by Google Chrome in early testing at the library, that hundreds of visitors are already set up with the capability to receive beacon proximity notifications. It is just a matter of showing them the value of recognizing and acting on these opportunities.

## Technical Hurdles

Although the web technology used was straightforward, dynamic webpages supported by a simple database, there was one unexpected hurdle resulting from Google Chrome's "network action predictions" (prefetch). When turned on by a cell phone user, this feature preloads any webpage that Chrome determines the user might visit ("Speed up Google Chrome," 2016). As the tracking for this project was done at the server level with PHP, if a visitor had both beacon support and prefetch enabled in the Chrome browser, the PHP tracking script saw Google preloading this page as a visit, even though the user had not actually viewed the page in his or her browser. At the time of this writing, Google does not identify a prefetched page in the HTTP headers sent to web servers, so there is no method to differentiate a real visit from a page preloaded by Google when a proximity beacon notification is generated by Chrome ("Issue 86175—chromium—Prerendering does not have any distinguishing HTTP headers," 2016).

During the initial phase of this pilot, the library saw hundreds of "false" visits to area pages, with only a visit to another beacon in the same session indicating the action was purposeful and not a prefetched page. Fortunately, client-side scripts are not rendered by the prefetch, so the JavaScript-based Google Analytics returns accurate statistics on visitors to the landing pages. This prefetch behavior should be a consideration for those implementing analytics on beacon visitors, as not only were local analytics misreported, but also the metrics provided by the PHY.net management tool showed a significant overreporting when compared to Google Analytics.

## Next Steps

While the pilot project revealed technical issues, it also revealed a great deal of potential. There are several other uses for proximity beacons in an academic library and on a university campus. The SDSU Library now has the opportunity to expand its two-year-old mobile tour by integrating proximity beacons. The current mobile tour largely features static content that is currently accessed through QR codes and NFC technology. The beacons offer a new method to engage students in exploring their campus library.

The beacon technology also offers an opportunity to expand the reach of research and instructional support beyond the areas of the library with service points. For instance, the government publication area of the library no longer has a dedicated service point where library users can get help with finding publications on the shelf. By leveraging proximity beacon technology, researchers can be connected to websites that feature information about how to read government publication call numbers, how to find items on the shelves, and how to contact librarians for help. Academic libraries have experimented with roving reference, a method of providing reference that involves a librarian roving through the stacks in order to find students who have research questions. While this method has offered an alternative to the traditional reference or research desk model, students who are unable to identify library staff have limited access to this service, which potentially interferes with student use of the library. Proximity beacons could be carried by library staff into the stacks, and any mobile devices in the immediate area would have the ability to connect and identify that someone is available to provide research assistance. This would mean that the student could contact the library staff member with the beacon, rather than having the staff interrupt the student to inquire about research assistance. While there are limitations to this method of providing reference or research services, it is an alternative some academic libraries may want to consider.

Proximity beacon technology can be used in a variety of ways, and one project the authors have proposed is a campus walking tour that would feature the art, architecture, and history of SDSU. This walking tour would work very similarly to the way the beacons have been used in the library, with websites being created for points of interest and the website URLs being pushed to mobile devices when they are close to a beacon. This tour would likely take the beacon project a step further by introducing sensors that will track that each beacon is operational. While the sensors were not needed for the small pilot project, a walking tour could feature twenty-five or more beacons, and sensors would make it easy to track each beacon's activity without having to check on each one individually. Each point of interest on this walking tour would feature content from the library's holdings, including images, videos, and links to the digital collections, Special Collections and University Archives, the catalog, and other interesting content that is relevant to the cultural or historical aspects of the university. The university has also expressed interest in enhancing campus wayfinding and has made inquiries into many aspects of this beacon pilot project to see if it can be adapted for campus wayfinding purposes. Additionally, the authors believe that the beacons can be used by the library to create interactive maps to help users navigate through the various areas of the two buildings.

More options exist for developing interactive projects with proximity beacons. One possibility to engage library users while gathering feedback would be to create a "voting wall" in the library. The wall could be something as simple as a flat-panel television that connects with the beacons and allows library users to answer questions on their mobile devices. Their answers would be recorded on the wall for everyone else to see. Libraries that are eager to implement low-interference methods of gathering user feedback could implement a proximity beacon–enabled voting wall quickly and at a relatively low cost.

There are many options for implementing proximity beacons in an academic library setting, and the cost of doing so can be fairly low. While the technology is fairly new, the implementation at the SDSU Library has shown promise. Librarians at SDSU will continue to experiment with the technology and encourage its adoption. There are certainly challenges in implementing the technology, but its adoption could lead to enhanced opportunities for academic libraries to connect with their users and provide them with access to relevant information.

# References

Aguirre, D., Johnston, B. & Kohn, L. (2011). QR codes go to college. Retrieved from https://web.archive.org/web/20111201011320/http://www.archrival.com/ideas/13/qr-codes-go-to-college.

BKON Connect, Inc. (2016). Beacons: 101: Simplified explanation of beacons. Retrieved from from http://bkon.com/blog/simple-explanation-of-ibeacon-and-eddystone-beacons/

Brown, M. (2014). Location, location, location. *Canadian Business, 87*(5/6), 12–14.

Faragher, R., & Harle, R. (2015). Location fingerprinting with bluetooth low energy beacons. *IEEE Journal on Selected Areas in Communications, 33*(11), 2418–2428. DOI: 10.1109/JSAC.2015.2430281.

Garun, N. (2014). Google launches Eddystone, its platform-agnostic iBeacon competitor. Retrieved from http://thenextweb.com/google/2015/07/14/google-launches-eddystone-its-platform-agnostic-ibeacon-competitor.

Gotttipati, H. (2013). With iBeacon, Apple is going to dump on NFC and embrace the internet of things. Retrieved from https://gigaom.com/2013/09/10/with-ibeacon-apple-is-going-to-dump-on-nfc-and-embrace-the-internet-of-things/.

Greenwald, A., Hampel, G., Phadke, C. & Poosala, V. (2011). An economically viable solution to geofencing for mass-market applications. *Bell Labs Technical Journal, 16*(2), 21–38. DOI: 10.1002/BLTJ.20500.

Higginbotham, S. (2015). Swirl raises $18M to tell the Internet where you are. Retrieved from http://fortune.com/2015/04/23/swirl-raises-18m-to-tell-the-internet-where-you-are/?iid=sr-link1.

Issue 86175—chromium—Prerendering does not have any distinguishing HTTP headers—Monorail. (2016). Retrieved from https://bugs.chromium.org/p/chromium/issues/detail?id=86175.

Jeffery, K., Jarocki, Z., & Baber, C. (2015). *Replacing the tour guide with a cell phone: A comparison of a guided and a self-guided library tour.* Poster session presented at the meeting of the American Library Association, San Francisco, CA.

Jenson, S., Want., R. & Schilit, B. (2015). Building an on-ramp for the internet of things. *IoT-Sys '15 Proceedings of the 2015 Workshop on IoT challenges in Mobile and Industrial Systems,* (pp. 3–6). DOI: 10.1145/2753476.2753483.

Michele, C. (2015). Apple iBeacons push 'proximity' marketing. *Investors Business Daily,* A09.

Monllos, K. (2016). 5 ways marketers are already putting sensors to work. *Adweek,* 1.

Phy.net. (2016). The importance of HTTPS for your Physical Web landing pages. Retrieved from https://www.phy.net/blog/http-vs-https/.

Speed up Google Chrome. (2016). Retrieved from https://support.google.com/chrome/answer/1385029?hl=en.

Want, R., Schilit, B. & Jenson, S. (2015). Enabling the internet of things. *Computer, 48*(1), 28–35. DOI: 10.1109/MC.2015.12.

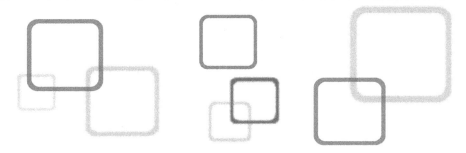

CHAPTER 14*

# Gamification Using Mobile Technology in the Classroom
## A Positive Benchmark for the Future of Higher Education

*Avery Le*

## Introduction

Technology can act as a double-edged sword, particularly in a classroom setting. As educators welcome the new generation of millennial students, they also hesitantly welcome the abundance of laptops and cell phones in the classroom. The digital age has propelled education into the twenty-first century and has altered traditional teaching methods. These devices connect students to an array of useful resources that can dramatically enhance one's learning experience. With the steady shift from print books to e-books, and from loose-leaf notebooks to Macbooks, librarians must readily adapt to the progression of this technological movement into academia as well. However, with this technology comes a source of distraction for students as the influx of entertainment applications and programs takes priority. How can instructors promote mobile technology as a powerful source of information instead? The solution is to not remove the entertainment

---

* This work is licensed under a Creative Commons Attribution- NonCommercial-NoDerivatives 4.0 License, CC BY-NC-ND (https://creativecommons.org/licenses/by-nc-nd/4.0/).

value from the technology, but rather to infuse it with educational themes that transform the classroom into an incubator for higher education. Instructors can redirect the students' focus by incorporating popular features of technology into the learning environment. Social media, games, chatting, e-mail, blogs—these digital trends have one common denominator—they keep users connected with each other. Users can enjoy each other's company virtually, at any point in time, and from anywhere in the world by logging into the same program.

Efforts at implementing gamification can capitalize on the proliferation of mobile devices by creating a synergy of interaction and learning that appeals to the modern-day student. Gamification has been defined as "the use of game design elements characteristic for games *in non-game contexts*, which is differentiated from playful design and a full-fledged game" (Kim, 2015, p.14). For the Lawton Chiles Legal Information Center, the "non-game context" occurred with librarians teaching a set of legal research competencies to first-year law students and assessing the students' ability to properly comprehend and apply these competencies. By framing this non-game context within the design elements present in a "game," instructors can take advantage of alternative learning styles.

## *Technology Simplifies the Implementation of Gamification*

Although cellular phones have existed as consumer devices since the early 90s, smartphones like the iPhone represented a new wave of mobility, combining the functionality of an Internet-capable computer with a practical design to emphasize the "mobile" factor in its light weight and sleekness. The iPhone was "about adapting the mobile to finally put it at the centre of computing, the Internet and digital culture" (Goggin, 2009, p. 232).

Marc Prensky fathered the term *digital natives* to describe the generation of students who innately incorporate the "digital language of computers, video games, and the Internet" into their daily lives (2001, p.1). *Digital natives* can barely recall a time when smartphones, tablets, and touch-screen laptops did not exist, as these devices are comparable to other household items, so it would be unrealistic to expect these students to relinquish the familiarity of instant gratification and immediate feedback often produced by the functions of these devices. Mobile technology invites a new level of personalization and customization, and libraries have had to adopt mobile technology to accommodate this customer-centric approach. An example of this is making a standard library website available as mobile version to accommodate various screen sizes and navigational differences (Bridges, Gascho, & Griggs, 2010, p. 313). Likewise, mobile-enabled teaching strategies cater to students in the classroom.

## Gamification Appeals to Law Students

Entry requirements to the law school include high GPAs and LSAT scores, which can be an indicator of high achievement amongst students. (University of Florida Levin College of Law, 2015). Indeed, gamification techniques were introduced in the law school regimen centuries ago, but without the label of *gamification* attached to it. Many law schools encourage the integration of practical skills in their curriculum, often represented by role-playing, as well as mock trials and negotiations such as Trial Team and Moot Court, and other nontraditional forms of gamification, such as Second Life (Wong, 2006). The face of gamification may have undergone several metamorphoses throughout the years, but the principles of gaming remain constant. With the new layer of mobile technology added, these principles exemplify a higher caliber of success. This transcendence of gamification fits the needs of law school students well, since it helps to provide the students with the appropriate skills to become competent and able attorneys.

# Gamification in the Classroom

As an effort to integrate mobile technology into the classroom, the author introduced two programs, Kahoot! and Socrative, to review sessions for Legal Research, a required one-credit course for first-year law students. These online programs powered quiz-type games consisting of multiple-choice questions that allowed the students to test their knowledge of the material they learned throughout the year. The game elements that prompted this decision were the prominent visualizations of the programs, extra motivation for the students to win, and the compatibility of the programs with the mobile devices that many students use daily.

The use of visualization as a way to attract the viewers' attention began with the introduction of PowerPoint slides as a lecturing tool (Hertz, Van Woerkum, & Kerkhof, 2015). However, it has become obvious that while visualizations might capture the students' attention momentarily, the images and patterns alone failed to connect the brain stimulation to the material, since graphical design and visualization have "little educational value unless they engage learners in an active learning activity" (Bouki, Economou & Kathrani, 2014, p.214). Programs such as Kahoot! are aesthetically pleasing, with its vibrant colors, cartoon-like animations, catchy sound effects, and interactive modes of competitive game play. Its effectiveness is accentuated not only through these visual and aural sensations but also through having the students perform a specific action in order to advance to the next stage in the game, something they cannot do without active engagement and participation.

Motivation is defined as "a relationship between feeling the need to experience a certain mood, the objective pursued and the possible means for achieving a certain goal" (Ejsing-Duun & Karoff, 2014, p.3). Motivation in games is achieved when a fair balance of skill and knowledge are presented but are not so challenging that the learning objective fades away. It is important to strike this balance because "achieving a sense of happiness is somewhere between the two [games that are too easy versus games that are too difficult]" (Parlic, Sofronijevic, & Cudanov, 2015, p. 139). With gamification, the syllabus expectations are converted into a goal: to acquire the most points in the game. This goal may lead to a prize, or at the very least, the self-satisfaction that comes with winning. The winner does not only win the "game," but also experiences the relief of knowing that he or she can comprehend the concepts and principles of legal research. Students who are not the identified winner benefit from the game as well, because it gives them a full analysis of the questions that they got wrong so that they can use them as a study guide.

When contemplating which mobile programs or applications best fit the classroom setting, the issue of compatibility with different mobile devices, operating systems, browsers, brand recognition, installation space, and Internet speed requirements are all important. The appeal of Kahoot! and Socrative is that they both require no additional installation, since they are web-based. Both programs are compatible with a variety of devices—Apple and Android smartphones and tablets, Windows and Mac operating systems, and popular browsers such as Safari, Internet Explorer, Chrome, and Firefox. This covers the gamut of mobile devices available for purchase and diminishes the possibility of the programs failing to run properly on a device.

## *Implementation of Kahoot!*

The Quiz option in Kahoot! was used to create the aforementioned games for the Legal Research review sessions. Kahoot! allows the students to play in real time against their classmates. The trivia game consists of a series of multiple-choice questions related to content given to the students through the textbook, lectures, PowerPoint slides, handouts, and class exercises during the semester. The objective of the game is to outscore the other players by answering the questions correctly and in the least amount of time. To join the game, the students simply enter a URL web address, www.kahoot.it, into a web browser using an Internet-enabled device of their choice, typically their laptops or smartphones. The website will prompt them for a five-digit PIN, which is provided by the instructor. After entering the PIN, the students must then enter a "nickname." The instructor can give them the option of creating an anonymous name if the game results do not count for a grade. There is no additional registration required, and no passwords, which alleviates any extended setup time or confusion. Questions are displayed on the administrator's screen only, meaning the players must be able to view the instructor's computer screen or projector. In order to score points, the students must read the multiple-choice question on the projector screen and answer correctly within the allotted time. Typically, the question will have four possible answer choices. With each second that goes by, the point value for the correct answer drops. Therefore, the faster a player answers the question correctly, the more points are awarded. Several students commented on a follow-up survey regarding their gaming experience in class that this intensity was not favorable to them, one student stating, "I think that made me pick the wrong answer because I felt too much pressure."

Another drawback of Kahoot! is that the text of the question and the answer choices appear on the main administrator screen (and projector) and the students can see only color-coded shapes to represent each answer choice on their mobile device screen (figure 14.1). This requires a careful balance of viewing both the projector at the front of the class and the mobile device screen to match them together. Once all of the players have answered the question or the time has expired, the correct answer is revealed on the projector's screen. After each question, the five players, with the most cumulative points are revealed on the scoreboard before the next question appears.

**FIGURE 14.1**
Screenshot of Kahoot! answer choices on mobile device.

## *Implementation of Socrative*

Socrative is very similar to Kahoot! with some notable differences in regard to pace, competitiveness level, and game play. Like Kahoot!, it uses a quiz-type game structure. However, in addition to multiple-choice questions, Socrative also allows for short-answer responses where students can type in their own free response. Socrative splits the game play up into "student-paced" versus "teacher-paced," which offers less competitive modes and concentrates more on the learning aspect. This program is more user-friendly in that both the question and the answer choices are clearly displayed on the mobile device's screen (figure 14.2).

**FIGURE 14.2**
Screenshot of questions on mobile device for Socrative.

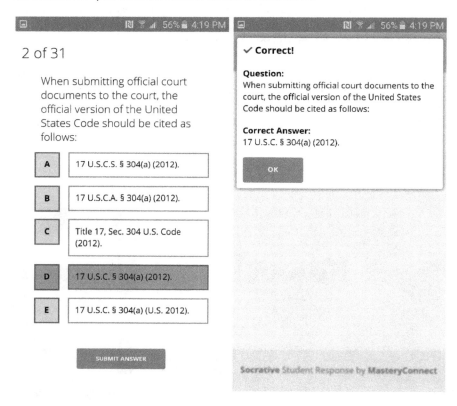

Within the teacher-paced mode, there is no timed factor such as that in Kahoot!, although the game will still display how many students have not locked in an answer yet. This mode also has the option to hide the percentage of students who selected the *wrong* answers, which minimizes room for the embarrassment of being singled out. Unlike in Kahoot!, there is no point differential based on how fast a player answers correctly. *However,* Socrative still offers a "Space Race" mode where students can be divided into teams either randomly or by their own choice. The method of answering is still conducted individually through each student's own mobile device. However, the scores are aggregated based on the team members' average. The scoreboard is displayed only at the end of the game and reveals the rankings of the teams, but not individual point values, making it more difficult to be critical of individual performances.

## The Benefits of Gamification
### *High Levels of Participation*
The features available in Kahoot! and Socrative enable an instructor to monitor the students' participation. In both programs every player can see the number of students who have already answered the question as well as the number of students still pondering the question. This added pressure forces students to focus entirely on the game and steers them away from other distractions such as their social media accounts, web surfing, or anything else unrelated to class—at least until they have finished answering a question. However, this configuration can cause apprehension for students knowing that the entire class may be waiting for them. Participation rate has become a vital part of the learning environment, and there is a positive correlation between active participation and comprehension of concepts. These games test the students' level of participation through visual, aural, and perceptive abilities all at once, and it is arguably more relaxed than the Socratic method.

### *Free Archive of Data and Analytics*
At the completion of the game, the results for each player are saved to the administrator's account. In Socrative, the results can be viewed for the class as a whole via an Excel spreadsheet, for individual players or questions as PDF files. There is another option to view the results as an online chart as well, without having to download any files to your computer. Access to the students' individual results derives from having each player answer the questions on separate devices as opposed to forming teams and sharing a device among the team members. Since the game moves forward at a quick pace, it can be burdensome for a student to keep track of the answers he or she selected without writing them down. However, electronic results are compiled throughout the game based on the players' answers, and the instructor can e-mail results to the students. One student commented, "I enjoyed being able to see where I was lacking knowledge and what topics I needed to dedicate more study time."

The results also help the instructor to recognize which questions were most problematic for the students. For example, if there was only a 50 percent accuracy rate for a question, the instructor can review it for wording issues, misleading information, clerical errors, or topics requiring more in-depth explanations. For the author, these analytics preempted any possible confusion with questions on the midterm and subsequently the questions on the exam.

### Individual Confidence Booster

The combination of individual performance and social interaction create a solid foundation for gamification that is both effective and impactful and gives millennial students their desired instant feedback (Suleman & Nelson, 2011). While traditional class exercises and group projects work well for students to bounce ideas off of each other, they can also result in one strong student taking the lead in the group, which can detract from the other students' learning process. On the other hand, the individual nature of games can provide students with a period of self-assessment to determine whether they truly understand the material. The independent structure can motivate students into taking careful notice of the difficult questions when no one else can answer for them, unlike during group work.

### Increased Motivation and Incentive

Once the students realized that the points were aggregated based on speed in Kahoot!, their level of interest skyrocketed. It was no longer about just getting the answer correct, but beating their classmates to the punch in selecting and entering the correct answer. The competition aspect pushed students to exert more effort than they normally would in exchange for a feeling of superiority at achieving a high score. The reward at stake, aside from bragging rights, was a "mystery prize." More than prize incentives, the motivation to achieve first place in the game gave the students a newfound sense of pride and the validation that their hard work in class benefited them. For example, one student stated in the survey feedback that it "tested knowledge in a way that made people pay attention and have some personal stake in it," and another stated that "the fast pace of Kahoot! challenged me and I think better prepared my brain for sensitive testing." Without the game's reinforcement, the students would not know whether their studying habits and preparation for the class were sufficient until after receiving their class grades, at which point there would be no chance to revise their approach. If the students did not do as well as they predicted in the game, they were motivated to spend more time reviewing their missed questions before taking the final examination.

## Possible Drawbacks

Possible drawbacks to gamification may cause instructors, particularly ones teaching legal research classes, to be hesitant to stray from their class routines. Some of these drawbacks are identified and discussed below.

## Overt Competitiveness

It can be argued that these games foster an overt competitiveness that can undermine the traditional learning environment and impose added stress to an already intense graduate program. Although law school is in itself built upon a foundation of competition, that competition culminates in the final examinations given to the students at the end of each semester, and only each student knows his or her class rankings unless he or she chooses to share that information with others. The only overt form of competition comes from the public listing of "Book Awards," accolades given to the best student in each course, typically the student who received the highest grade in the course. Here, the Kahoot! and Socrative games do crown immediate winners, and the objective is clearly defined at the start of the game. The overtness of pitting the students against one another may cause anxiety and stress for some students.

## Students with Learning Disabilities

In Kahoot!, having to match shapes with a corresponding answer choices may become problematic for students with learning disabilities. For the majority of students, this is not a burdensome task, as memory and association are pivotal skills that many students have acquired by the time they reach graduate school. However, the ability to recognize shapes and colors may be difficult for students with learning disabilities that affect visual processing and spatial recognition. On the other hand, according to Brown and colleagues, there have been several studies that suggest that gamification positively motivates students with learning disabilities if it is properly designed and implemented with these students in mind and can generate a novel way of delivering information to them through game-like elements as it diffuses the rigidity and strenuous pressures of testing environments (Brown et al., 2013). They note that "research has shown that [Digital Game-Based Learning] can have a positive effect on some of the core development needs of people with Intellectual Disabilities and associated sensory impairments" (Brown et al., 2013, p.574).

There are various stress factors that can affect registered ADA (Accessibility and Disability Accommodations) students. For instance, some ADA students do not thrive in high-pressure situations, especially in group settings. These students often keep their identities hidden during class to avoid any backlash for receiving accommodations for quizzes and tests. These games simulate a quiz because they involve answering multiple-choice questions correctly. ADA students are often given more time for quizzes and tests, so the pressure of participating in these games may put them at a disadvantage, even though the games may not count toward their grade. This pressure may have a negative impact on students if they struggle to keep up or find themselves repeatedly missing questions during the game due to the time constraints.

## *Disadvantage for Non-Tech-Savvy Students*

Members of the eighteen-to-twenty-nine-year-old age group are mostly familiar with and enjoy playing video games, so it is not a surprise that this age group is most open to gamification efforts (Duggan, 2015). This age group also coincides with the age range of incoming student enrollment at the University of Florida Levin College of Law. In 2015, students ranged in age from eighteen to fifty-one years old and the majority fell in the middle of the range, with outliers in the forty and above subsection (University of Florida Levin College of Law, 2015). The students who fall out of this particular millennial age range, those who are not digital natives, and millennial students who do not regard video games as a favorite pastime, may feel out of sorts with gamification that replicates commonalities of video games such as testing agility and speed through rapid button pressing and quick responses to score points. It may even be a student's first time playing an interactive game against other players. The non-tech-savvy students may regard the games as an obstacle in their educational progress.

## *Exclusion of Students without a Mobile Device*

The use of gamification coupled with technological devices also poses a possible issue for those who lack the necessary mobile device and therefore cannot participate. However, the University of Florida Levin College of Law has a mandatory laptop requirement for all enrolled students to support its emphasis on preparing its students "to be technologically sophisticated in the use of computers and computerized legal research" (University of Florida Levin College of Law, n.d.). As most students will come equipped with at least one mobile device, a laptop, the use of gamification through mobile technology has not posed an issue for the author's students. While there is a general presumption that all law students will comply with the technology requirement and have the financial means to do so, this is not a reality for students who carry a major financial burden. Therefore, implementing these games within the classroom may feel unfair if they cannot participate without a mobile device.

## *Technical Difficulties*

Technical difficulties present another possible disruption to gamification, especially with the programs like Kahoot! and Socrative that require a high-speed Internet connection. One student complained that his fourth-generation iPhone could not keep up with the game and that the Internet lag caused a delay in his game play. Another mishap can occur from a player accidentally exiting the browser. Rejoining the game is possible, but it will pick up from where the rest of the players are, so any previous questions left unanswered will be considered incorrect if the student joined the game later. However, these were rare occurrences. The author therefore recommends that these games exist as a supplemental tool for the classroom and not as the main component of every lesson to ameliorate such problems.

## Solutions for Drawbacks

There are ways to mitigate the possible negative aspects in order to minimize any post-gamification detriments. For example, the instructor can select less competitive modes that do not stress point values and high scores. There are modes, available through Socrative, that do not require any points to be assigned to the questions. The students can also be divided into teams so that no one is singled out for any wrong answers. In Socrative, the instructor can hide the results that reveal what percentage of the class chose what answer so that no one is embarrassed over choosing any obviously incorrect answers. The instructor can also assign more time to each question to give each student a fair chance of answering correctly or to read the question more carefully. Gaming can be introduced gradually by conducting a practice run to give students a chance to familiarize themselves with the programs. One student noted that although "the game was great, I would've enjoyed it more if I knew I would be playing it and had time to prepare." Some of the librarians used Kahoot! as an icebreaker during orientation, which is a preliminary class session prior to the first week of school, to "test" the students' knowledge of library policies. This allowed for the game to be introduced to the students early so that they already knew what the rules and instructions were and how to accumulate points by the time the games were reintroduced later in the semester. The setup time is then much quicker, and less explanation is needed on the part of the instructor. It is recommended that instructors research which mobile program best complements the personalities and needs of their students before committing to gamification.

## Conclusion

Overall, the positive attributes outweigh the negative in regard to the implementation of gamification through mobile technology. The majority of students enjoyed competing against others as a way to review material from the class. Moreover, gamification offers students necessary mental relief from the intensity of law school and reinforces a collegial classroom dynamic. The social interaction strengthens the students' relationships with each other as well as with the professor and provides a level of comfort and builds rapport that results in the students working well together on group projects and allows the students to more easily approach the professor with their questions.

The intricacies of mobile technology allow for games to be hosted on a sophisticated, structured, and aesthetically pleasing electronic platform. The advantage to integrating these games into mobile technology to create real-time live interaction is that no additional hardware or accessories are necessary. Other computer-based trivia games have been marketed to instructors but have required devices such as costly high-end wireless buzzers, with a need to replace batteries or replace those that malfunction or wear out ("Who's First Wireless Game Show Buzzer System"). The facts that Kahoot! and Socrative are currently free to everyone and that they are compatible with devices that the students already own make them appealing.

As technological advances continue, gamification will only become more refined. We are constantly seeing a new wave of technology emerge, such as virtual reality add-ons to smartphones, the Internet of Things, and hands-free game play powered by video game consoles such as the Xbox and PlayStation Move, or HoloLens—the "first fully untethered, holographic computer," that will power hologram applications currently in development by Microsoft Windows ("Microsoft", 2016). It is difficult to predict which gaming applications will remain at the forefront of the industry. For now, programs such as Kahoot! and Socrative prove to be solid extensions to the classroom and will continue to increase the confidence of law students and simulate pressure-packed situations and to help develop social skills that these students will need to succeed in their legal careers.

# References

Bouki, V., Economou, D., & Kathrani, P. (2014, 13–14 November). *"Gamification" and legal education: A game based application for teaching university law students.* Paper presented at the 2014 International Conference on Interactive Mobile Communication Technologies and Learning (IMCL). Retrieved May 1, 2016, from IEEE Xplore.

Brown, D., Standen, P., Saridaki, M., Shopland, N., Roinioti, E., Evett, L. Grantham, S., & Smith, P. (2013). Engaging students with intellectual disabilities through games based learning and related technologies. In Constantine Stephanidis & Margherita Antona (Eds.), *universal access in human-computer interaction, applications and services for quality of life*. Berlin: Springer.

Bridges, L., Gascho, H., & Griggs, K. (2010). Making the case for a fully mobile library web site: From floor maps to the catalog. *Reference Services Review*, 38(2), 309–20. doi:10.1108/00907321011045061.

Duggan, Maeve. (2015) "Gaming and Gamers". Pew Research Center. Retrieved from http://www.pewinternet.org/files/2015/12/PI_2015-12-15_gaming-and-gamers_FINAL.pdf.

Ejsing-Duun, S. & Skovbjerg, H. (2014). "Gamification of a higher education course: What's the fun in that?", Paper read at ECGBL2014-8th European Conference on Games Based Learning, Berlin, Germany.

Goggin, G. (2009). Adapting the mobile phone: The iPhone and its consumption. *Continuum: Journal of Media & Cultural Studies*, 23(2), 231–44.

Hertz, B., Van Woerkum, C., & Kerkhof, P. (2015). Why do scholars use powerpoint the way they do? *Business and Professional Communication Quarterly, 78*(3), 273–291. doi:10.1177/2329490615589171.

Prensky, M. (2001). "Digital natives, digital immigrants". In *On the horizon, October 2001, 9*(5). Lincoln: NCB University Press.

Kim, B. (2015). Understanding gamification. *Library Technology Reports, 51*(2), 1–35. Web. 29 Apr. 2016.

Microsoft. (2016). *Microsoft Windows HoloLens*. Retrieved from http://www.microsoft.com/microsoft-hololens.

Suleman, R., & Nelson, B. (2011). Motivating the millennials: Tapping into the potential of the youngest generation. *Leader to Leader, 62*, 39–44. doi:10.1002/ltl.491.

Toyama, K. (2015, October 29). The looming gamification of higher ed. *The Chronicle of Higher Education*. Retrieved from http://www.chronicle.com.

University of Florida Levin College of Law. (2015). *Florida's flagship*. Retrieved from https://www.law.ufl.edu/_pdf/admissions/prospective/viewbook-2015-2016.pdf

University of Florida Levin College of Law. (n.d.) *Mandatory laptop requirement*. Retrieved from https://www.law.ufl.edu/why-uf-law/about-uf-law/services/technology-services/mandatory-laptop-requirement.

*Who's first wireless game show buzz system*. (n.d.). Retrieved from http://www.training-games.com/whos-first-wireless-buzzers/.

Wong, G. (2006, November 14). Educators explore 'Second Life' online. *Cable News Network (CNN)*. Retrieved from http://www.cnn.com.

CHAPTER 15*

# Bringing Texts to Life
## An Augmented Reality Application for Supporting the Development of Information Literacy Skills

*Yusuke Ishimura and Martin Masek*

## The Undergraduate Student's Experience of Writing Research Papers

Library and information science (LIS) research has frequently discussed undergraduate students' difficulties in conducting independent research. Leckie's (1996) notable study illustrated that faculty members' assumptions about their students' research processes often lead to difficulty for the students in completing their research assignments. Many studies confirm that it is not an easy task for undergraduates to conduct independent research and to write academic papers. For example, Head and Eisenberg (2010) found that the majority of students they studied in large academic universities in the United States had difficulties in initiating the research process. Many undergraduate students do not have confidence completing assigned tasks and have anxiety when they first start (Dubicki, 2015). Detmering and Johnson's (2012) study of undergraduate students discovered that students often portray themselves as "heroes facing the challenge of writing a research paper" with challenging restrictions and rules set by their instructors (p. 11).

---

* This work is licensed under a Creative Commons Attribution- NonCommercial-NoDerivatives 4.0 License, CC BY-NC-ND (https://creativecommons.org/licenses/by-nc-nd/4.0/).

Kuhlthau's (1991, 1993, 1995, 2004, 2008) well-known information-seeking behavior model, Information Search Process (ISP), highlighted that students feel the highest anxiety during the exploration stage of their process and experience difficulties completing assigned tasks independently. However, it can be challenging for educators and librarians to provide "just in time" assistance to students. In many cases, educators are not in contact with students while they are doing independent research tasks. How can the learning experience of students be supported when individualized expert advice is not available? This chapter attempts to answer this question by using a cutting-edge technology, augmented reality (AR).

The overall objective of this study is to investigate how AR technology can support on- and off-campus students during the research paper writing process. More specifically, this study develops an AR mobile application that allows students to be able to interact with digital layers of various multimedia superimposed on academic sources as a preparation for completing a writing assignment. The study also investigates how students perform various tasks using the AR application.

## What Is Augmented Reality?

In the *Horizon Report* (New Media Consortium, 2016), AR is predicted to be a technology that will be adopted within two to three years in higher education. Azuma and colleagues (2001) explain that AR systems display virtual (computer-generated) objects in the real world, and the objects and the physical world coexist in the same space. They define AR as a system that "combines real and virtual objects in a real environment; runs interactively, and in real time; and registers (aligns) real and virtual objects with each other" (p. 34). AR is a technology that contextualizes real objects, or places, by adding information and meaning to them so that users can understand the physical items in greater depth. AR technology can overlay images, add audio commentary, pull in location data, or show historical information or other relevant content for users to interact with objects or places (EDUCAUSE, 2005). Simply put, the technology can enhance our physical world and promote exploration by adding digital layers of information over various real objects.

AR technology is becoming more widely used for teaching and learning in various educational settings, and we'll begin by describing a few representative applications of the technology. By scanning a textbook with special markers, elementary school students using an AR application will be able to engage with content such as 3-D objects for learning about geometry (Dionisio Correa et al., 2013). Another example is students studying forensic medicine who can scan predefined markers on their body (e.g., neck or arm) using an iPhone to be able to see multimedia content in the AR application explaining wound patterns (Albrecht, Folta-Schoofs, Behrends, & von Jan, 2013). Using multiple AR applications, electrical engineering students can learn step-by-step instructions on how to use electrical machines, translate electrical symbols into images, and view a study "notebook" to understand theoretical content that is often difficult to understand without looking at 3-D shapes (Martín-Gutiérrez, Fabiani, Benesova, Meneses, & Mora, 2015).

The use of AR in education is not limited to formal courses, and there are a few examples in the literature of academic libraries using the technology. For example, Barnes and Brammer (2013) discussed the potential use of the Stiktu application for a library scavenger hunt. The application can be used for solving problems related to the specific resources scanned or for helping students become more familiar with the library building and collection. Another example is the use of AR to provide assistance when students are browsing library stacks, providing information about topics such as the sections they are browsing, past circulation counts of a specific item, and recommended resources related to items scanned with the application (Hahn, 2012). In the school library setting, Mulch (2014) explained an AR library tour initiative in which students created the content of the tour. When students scan targets in the library, videos appear in the app for them to engage with. Although a few initiatives such as these exist, the application of AR is not yet widely discussed in the LIS field. Moreover, the application of AR technology from an information literacy perspective is an emerging area of interest that has not been researched to date.

# Development of the AR Application: The Trailblazer Project and Text Recognition

The Trailblazer text recognition project was initiated in January 2014 with support from an Edith Cowan University, Faculty of Health, Engineering, and Science Teaching and Learning grant. The purpose of the grant award focuses on improving students' learning experience in the university.

Most AR applications use large-scale icons or barcode-like patterns to anchor virtual content in the real world. For the application in this project, however, the augmented content needed to appear linked to paragraphs and passages of text in a document, which is a two-dimensional object. To support this unusual functionality, the authors modified an existing AR project, Trailblazer (Boston, Masek, Brogan, & Lam, 2014). Trailblazer is an AR application developed in partnership between Edith Cowan University and the National Trust of Australia in Western Australia. Using mobile phones and tablet devices that run on the Android OS, the AR application's users interact with virtual content based on either geolocation or visual markers. The content includes overlays of text, images, videos, 3-D objects, and interactive quizzes. The image-based AR capability of Trailblazer is provided by PTC Vuforia, an application-authoring software.

To enable the use of AR overlays on sections of document texts for the project described here, a process needed to be developed for visual marker creation. The relevant page of a PDF document was opened in an image-editing package, such as the freely available GNU Image Manipulation Program (GIMP), with at least 800 pixels per inch resolution. The contrast of the page was then enhanced, also through the image editor, using the Levels adjustment tool. Through this tool, the color histogram (that is, the distribution of colors) of the page was used to visually place a lower and upper bound on pixel intensity from the original version of the page. After that, the pixel intensities could be remapped between this narrowed range to the full range of the image format. This maximized the contrast of the text while avoiding external interference. This contrast enhancement assisted Vuforia's feature detection algorithm and allowed for higher reliability in detection and tracking. In other words, the processing made the black-and-white text recognizable to the software. Each page of the PDF document was processed this way and then used to create a database of markers, or visual patterns that could be recognized by Vuforia, using Vuforia's online creation tool. This database was then imported into Trailblazer.

To maximize marker detection robustness, the application was developed to use each entire page as a marker, as the more geometric detail present in the marker image, the better the detection probability. Although to the human eye, pages of text may look very similar from a distance, the high resolution of current mobile device cameras allows for tracking of the minutiae of corners and edges of individual letters in the text. The relative distances between these features form a unique signature for each page. Figure 15.1 shows a zoomed-in example of a section of the page with features that are used by Vuforia to recognize the text indicated.

**FIGURE 15.1**
Example of features used to recognize sections of text. The markers overlaid on the text indicate the sections used by Vuforia to recognize the page of text.

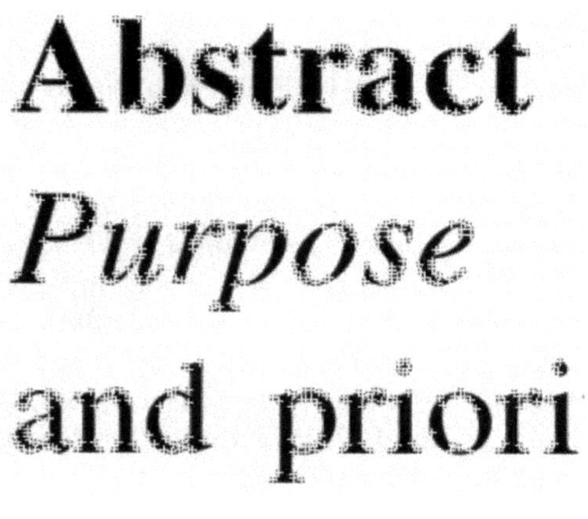

**Abstract**

*Purpose*

and priori

Once a page has been successfully identified, our goal was to overlay a specific portion of it, rather than the whole page, with augmented content designed to assist students in their research work by helping them to understand the basic structure and organization of an academic article. To allow for this, an overlay editor tool was developed, as shown in the screenshot in figure 15.2. The tool allows the application user to define an area visually and to define what type of content is assigned to it. The output of this application is an XML file that was loaded into Traiblazer.

### FIGURE 15.2
The Overlay application was used to position content on specific sections of a document, in this example the location and size of an info panel.

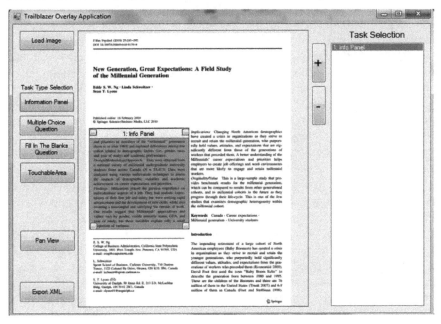

Text recognition is crucial for the application development. Through this process, the authors successfully programmed the application to recognize textual information. The next step was to develop the digital content to overlay on the article.

## Content Development

Writing a research paper is often a long and daunting process for undergraduate students. To complete the paper successfully, students need to understand how the research process works, as discussed in the Kuhlthau's (2004) ISP model. In addition to the process, they should understand the context of the process as a part of their learning. Students need to have skills in interacting with information and integrating their new learning into their current knowledge. However, as the Association of College and Research Libraries' (ACRL, 2015) recent information literacy framework makes clear, utilizing information effectively and developing information literacy are not simply a generic set of skills, but contextual ways of thinking and acting. It is also necessary for students to understand how information is created and valued. This project focused particularly on the initial phase of students' research process, namely the understanding of basic information literacy concepts and the nature of academic journal articles, as preparation for later steps of assignment tasks. Thus, the first version of the mobile application was designed specifically to help students in an undergraduate management course targeting computer science students to understand basic concepts about the content and structure of an academic article. With this context provided, it is hoped that students will be better able to develop the potential direction of their research, seek external sources, and write a research paper.

In order to develop the content for the AR application, the authors first framed various activities based on reading comprehension theories. Many computer science students at the university are not native speakers of English, and so a particular focus was placed on theories of reading English as a second language (Grabe, 2009). As Grabe pointed out, if students cannot engage in and understand the content, they will not be able to complete their assigned tasks effectively. Reading is not simply skimming text. It is important for students to interact with texts and understand the content. Grabe further claimed that successful readers have higher metacognitive knowledge than novices, so students need to develop reading strategies and monitor comprehension levels. The application was designed to help students learn these steps.

Another layer of the application development is the integration of information literacy elements into activities. Keeping ACRL's (2015) *Framework for Information Literacy* in mind, students need to understand the basic bibliographic information, structure of the journal article, and style of writing in scholarly publications before moving on to a more sophisticated understanding of the material. The first step was to select a research article for this project with which to pilot the approach. In the first instance, the authors selected just one article in order to store information about it in the application. A journal article was sought that covers a management topic that is relevant and interesting for computer science students. The following item was selected:

> Ng, E. S. W., Schweitzer, L., & Lyons, S. T. (2010). New generation, great expectations: A field study of the Millennial generation. *Journal of Business and Psychology*, 25(2), 281–292.

After the paper was selected, AR activities were developed that would support students in the early stages of their independent writing assignment. As table 15.1 shows, activities were divided in three stages: introduction, pre-reading, and during reading. We first read the selected paper for the pilot testing and identified important elements and ideas according to the goals of the activities that would be assigned to the students. Next, questions were created in accordance with the reading stages. Although more activities could be added, this project was limited to ten items so that the activities could be completed within twenty to thirty minutes. The application development system provides different types of questions (i.e., short answer, fill in the blank, and multiple choices) that can be selected.

**TABLE 15.1**
Overview of augmented reality activities.

| Reading stages | Purpose | Activities | Question types |
|---|---|---|---|
| Introduction | Overall instruction & explanation of tasks | — | — |
| Pre-reading | Understanding of basic bibliographic information | Text explanation: Connecting course content with the paper | — |
| | | Q1: Identifying journal volume number | Short answer |
| During reading | Understanding paper content | Q2: Understand content in abstract | Fill in the blank |
| | | Text explanation: Purpose of introduction | — |
| | | Q3: Identify the purpose of paper from introduction section | Multiple choice |
| | | Text explanation: Conclusion section | — |
| | | Q4: Identify millennial generations' expectations as argued in the article | Multiple choice |
| | | Text explanation: Literature review section | — |
| | | Q5: Good pay and benefits as argued in the article | Fill in the blank |
| | | Text explanation: Method section | — |

**TABLE 15.1**
Overview of augmented reality activities.

| Reading stages | Purpose | Activities | Question types |
|---|---|---|---|
| During reading | Understanding paper content | Q6: Identifying attributes related to selecting jobs as argued in the article | Short answer |
| | | Text explanation: Results section | — |
| | | Q7: Identifying impact on promotion expectations as argued in the article | Multiple choice |
| | | Q8: Identifying differences between non-minority and minority groups as argued in the article | Multiple choice |
| | | Text explanation: Discussion section | — |
| | | Q9: Implications to women's status in labor market as argued in the article | Fill in a blank |
| | | Q10: How to attract students with high GPA as argued in the article | Multiple choice |

Figure 15.3 shows the actual application interface. When students scan the assigned reading (on either screen or paper), digital information shows up on the screen relevant to the page they are currently viewing. In addition to a "quiz" question, a target marker displays on the screen to tell students where to look for information to answer the question.

# Results
## Analysis of Students' Performance

Once the prototype application and activities were developed, a testing phase was entered with students. In order to test how students interacted with the application, the authors asked them to complete all tasks on tablet devices that were provided to them. This project underwent a human research ethics review before the testing phase, and students gave informed consent to participate through an agreement form on the testing devices.

## FIGURE 15.3
Screenshot of augmented reality application prototype.

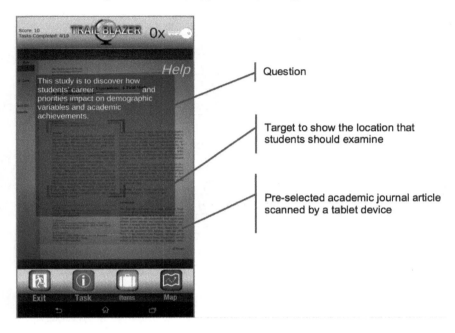

All participants in the testing ($n = 12$) were undergraduate computer science students taking CSG3204: Information Services Management at Edith Cowan University. The testing was conducted during one class period (approximately sixty minutes). The application on the tablet devices is capable of collecting various data on each question as a CSV (comma-separated values) file. At the end of all tasks, students are able to send the results to instructors, or instructors can retrieve the results stored in the tablet. Table 15.2 summarizes students' performance in scores (10 points maximum, and points reduced if students fail), time spent on each question, and number of false attempts. By analyzing the performance data, it was possible to tell which questions were difficult to answer (i.e., questions 3 and 7), which questions took more time to answer (i.e., questions 3, 5, and 6), and which questions students made mistakes on most often (i.e., questions 3, 6, and 8). This functionality demonstrates the potential of analytics data collected by mobile devices to enable instructors to investigate students' performance more deeply and modify their teaching approach in response to this.

**TABLE 15.2**
Summary of scores, time spent, and false attempts (n = 12).

| Question | Score (10 points max.) | | Time spent (seconds) | | # of false attempts | |
|---|---|---|---|---|---|---|
| | Mean | Std. Dev. | Mean | Std. Dev. | Mean | Std. Dev. |
| Q1 | 9.23 | 2.77 | 64.80 | 29.37 | 0.15 | 0.55 |
| Q2 | 10.00 | 0.00 | 38.62 | 20.12 | 0.00 | 0.00 |
| Q3 | 3.85 | 5.06 | 105.23 | 54.32 | 1.08 | 1.04 |
| Q4 | 5.38 | 5.19 | 60.81 | 29.61 | 0.85 | 1.41 |
| Q5 | 7.69 | 4.39 | 164.09 | 76.15 | 0.38 | 0.77 |
| Q6 | 4.62 | 5.19 | 102.93 | 102.77 | 1.85 | 2.44 |
| Q7 | 3.08 | 4.80 | 67.76 | 61.29 | 0.69 | 0.48 |
| Q8 | 6.92 | 4.80 | 77.92 | 63.59 | 1.08 | 2.53 |
| Q9 | 10.00 | 0.00 | 57.28 | 27.82 | 0.00 | 0.00 |
| Q10 | 5.38 | 5.19 | 45.37 | 28.63 | 0.46 | 0.52 |
| All questions | 66.15 | 15.56 | 78.48 | 64.17 | 0.65 | 1.37 |

## *Testing Application Usability*

In addition to students' performance on tablet devices, the study also investigated whether our application prototype is easy to use. Using the System Usability Scale (SUS) (Brooke, 1996), the application was tested after students' use of the tablet devices (see table 15.3). Although this scale was developed in 1996, it is still considered valid to test usability (Lewis & Sauro, 2009). The SUS contains 5-point Likert scale questions. Even-numbered items are scored on a range from 0 to 10 points, and odd-numbered items range from 10 to 0. This questionnaire has ten questions, and therefore a full score for usability would be 100. The mean total score of our application was 67.71. SUS's average score for programs is 68 points. This suggests that our prototype has a passing grade in terms of usability. More details of future improvements in response to this testing are discussed in the next section.

**TABLE 15.3**
System Usability Scale results.

| Usability questions | N | Mean | Std. Dev. |
|---|---|---|---|
| Q1: I think that I would like to use this app frequently. | 12 | 5.63 | 1.88 |
| Q2: I found the app unnecessarily complex. | 12 | 6.46 | 1.98 |
| Q3: I thought the app was easy to use. | 12 | 6.67 | 2.22 |

| TABLE 15.3 System Usability Scale results. | | | |
|---|---|---|---|
| Usability questions | N | Mean | Std. Dev. |
| Q4. I think that I would need the support of a technical person to be able to use this app. | 12 | 8.75 | 2.26 |
| Q5. I found the various functions in this app were well integrated. | 12 | 6.67 | 1.23 |
| Q6. I thought there was too much inconsistency in this app. | 12 | 6.88 | 2.17 |
| Q7. I would imagine that most people would learn to use this app very quickly. | 12 | 7.50 | 1.85 |
| Q8. I found the app very cumbersome to use. | 12 | 4.17 | 2.46 |
| Q9. I felt very confident using the app. | 12 | 7.08 | 2.34 |
| Q10. I needed to learn a lot of things before I could get going with this app. | 12 | 7.92 | 2.98 |
| Total score | 12 | 67.71 | 12.22 |

## *Qualitative Feedback*

Qualitative feedback from students revealed that they were impressed with the application's capabilities. In particular, they were intrigued by the fact that simple black-and-white textual information can become interactive. Also, the students were impressed by how quickly the application recognizes text and pages. They described this as "cool." This is a very important description for the researchers because it helps to demonstrate the usefulness of this application's development, as the application is meant to be engaging and helpful, especially in the early stages of an assignment process. Another point mentioned by the students surveyed was the target audience. Since many participants were final-year students, they recommended that the application would be useful for first-year undergraduate students. Based on feedback from students, some areas emerged to suggest improvements in developing the application. Some items have already been changed, while others will be modified in the near future.

### INTERFACE DEVELOPMENT

Some students pointed out that the user interface was a bit dated looking and needed to be changed. Considering that the application was designed for computer science students, the authors were concerned that it would not be attractive to those who often interact with cutting-edge technologies. The interface has now been made "flat" looking instead of using shadows around buttons and objects, consistent with recent trends in mobile apps and web design. It is hoped that this updated design can encourage more frequent use of the app by students.

### DISTANCE TO PAPER

Students need to fit entire pages into the application using cameras on tablet devices in order for the digital content to appear. This forces them to keep tablets at a certain distance. Until students get used to the distance required, it is challenging to do so. Also, they need to hold the tablets still until they complete activities because continuous scanning is required to view and complete activities. In order to improve on this deficiency, the application's processing will be changed so that students do not have to hold tablet devices at a certain location. More specifically, when they scan a paper going forward, the activities will stay on the screen even if they move the devices away from the paper. Also, it is possible to capture screenshots so that they can see a marker and a specific page on the screen even if they have moved the tablet, a potential new feature that will be explored in future development.

### PORTRAIT MODE

Based on our observation and feedback comments, the application should be locked to portrait mode. A couple of participants tried to use the device in landscape mode, and they were frustrated with that orientation. Since a journal article is traditionally in portrait orientation, the application was designed to be used in that orientation. The ability to lock the app in portrait mode has now been implemented.

### GAMING ASPECTS

Some students informally suggested that the application could be modified to include a game aspect. For example, one student suggested that they could use the application with an avatar, and that when they make mistakes, their character would lose a life. Then, when they use up all of their avatar's lives, the game would be over. As gamification is an emerging subject of interest in learning and technology, the integration of gaming elements into the application will be investigated for future development.

## Conclusion and Future Direction

This project began as an experiment, and our experiences suggest many interesting possibilities for the future use and development of augmented reality apps. The authors faced challenges in making the application recognize text as opposed to a location or physical visual markers as is usually done with AR applications.

However, despite the hurdles, and after a few iterations, the application was successfully able to recognize textual information. Our initiative proves that the use of the AR technology is not limited to location and visual objects. As discussed, the possibility of using AR to support information literacy development has not yet been fully explored. This application has the potential to add to our knowledge and practice in information literacy, and LIS in general, in demonstrating how AR can be used to support students in particular stages of their assignment process. It is hoped that this study will stimulate more active development of AR applications to improve university students' learning, especially in information literacy skills development.

Tracking student learning is another potential benefit of using this application. Educators are not always available when students engage in learning. The application is developed as a mobile application and is capable of storing performance data. The recording capability will enable educators to make certain that students meet the learning objectives of an assignment, and also to better adapt their teaching methods to students' needs.

The project described here was an exploratory study to test the capability of the technology. The effectiveness of the application against specific learning outcomes has not yet been tested. Future investigations will include conducting a comparative study between control and experimental groups using different learning modes.

# Acknowledgment

We would like to acknowledge Amos Wolfe, who developed this application as our research assistant. We also would like to thank the Edith Cowan University, Faculty of Health, Engineering and Science, for the Learning and Teaching grant awarded to support this project.

# References

Albrecht, U.-V., Folta-Schoofs, K., Behrends, M., & von Jan, U. (2013). Effects of mobile augmented reality learning compared to textbook learning on medical students: Randomized controlled pilot study. *Journal of Medical Internet Research*, 15(8). http://doi.org/10.2196/jmir.2497.

Association of College and Research Libraries. (2015). Framework for information literacy for higher education. Retrieved from http://www.ala.org/acrl/standards/ilframework.

Azuma, R., Baillot, Y., Behringer, R., Feiner, S., Julier, S., & MacIntyre, B. (2001). Recent advances in augmented reality. *IEEE Computer Graphics and Applications*, 21(6), 34–47. http://doi.org/10.1109/38.963459.

Barnes, E., & Brammer, R. M. (2013). Bringing augmented reality to the academic law library: Our experiences with an augmented reality app. *AALL Spectrum*, 17(4), 13–15.

Boston, J., Masek, M., Brogan, M., & Lam, P. (2014). *Learning in digitally augmented physical spaces*. Proceedings of the Australian Computers in Education Conference, 52–60. Retrieved from http://acec2014.acce.edu.au/sites/2014/files/2014ConfProceedingsFinal.pdf.

Brooke, J. (1996). SUS: A 'quick and dirty' usability scale. In P. W. Jordan, B. Thomas, B. A. Weerdmeester, & I. L. McClelland (Eds.), *Usability evaluation in industry* (pp. 189–194). London, England: Taylor & Francis.

Detmering, R., & Johnson, A. M. (2012). Research papers have always seemed very daunting: Information literacy narratives and the student research experience. *Portal: Libraries and the Academy, 12*(1), 5–22. http://doi.org/10.1353/pla.2012.0004.

Dionisio Correa, A. G., Tahira, A., Ribeir, J. B., Kitamura, R. K., Inoue, T. Y., & Karaguilla Ficheman, I. (2013). Development of an interactive book with Augmented Reality for mobile learning. *Proceedings of the Iberian Conference on Information Systems and Technologies*, 1–7.

Dubicki, E. (2015). Writing a research paper: Students explain their process. *Reference Services Review, 43*(4), 673–688. https://doi.org/10.1108/RSR-07-2015-0036.

EDUCAUSE. (2005, October 15). *7 things you should know about augmented reality*. Retrieved from https://library.educause.edu/~/media/files/library/2005/10/eli7007-pdf.pdf.

Grabe, W. (2009). *Reading in a second language: Moving from theory to practice*. New York, NY: Cambridge University Press.

Hahn, J. (2012). Mobile augmented reality applications for library services. *New Library World, 113*(9/10), 429–438. http://doi.org/10.1108/03074801211273902.

Head, A. J., & Eisenberg, M. (2010). *Truth be told: How college students evaluate and use information in the digital age*. Retrieved from http://www.projectinfolit.org/uploads/2/7/5/4/27541717/pil_fall2010_survey_fullreport1.pdf.

Kuhlthau, C. C. (1991). Inside the search process: Information seeking from the user's perspective. *Journal of the American Society for Information Science, 42*(5), 361–371.

Kuhlthau, C. C. (1993). A principle of uncertainty for information seeking. *Journal of Documentation, 49*(4), 339–355.

Kuhlthau, C. C. (1995). The process of learning from information. *School Libraries Worldwide, 1*(1), 1–12.

Kuhlthau, C. C. (2004). *Seeking meaning: A process approach to library and information services* (2nd ed.). Norwood, NJ: Ablex.

Kuhlthau, C. C. (2008). From information to meaning: Confronting challenges of the twenty-first century. *Libri, 58*(2), 66–73.

Leckie, G. J. (1996). Desperately seeking citations: Uncovering faculty assumptions about the undergraduate research process. *The Journal of Academic Librarianship, 22*(3), 201–208. http://doi.org/10.1016/S0099-1333(96)90059-2.

Lewis, J. R., & Sauro, J. (2009). The factor structure of the system usability scale. In M. Kurosu (Ed.), *Human Centered Design* (Vol. 5619, pp. 94–103). Berlin, Germany: Springer Berlin Heidelberg. Retrieved from http://doi.org/10.1007/978-3-642-02806-9_12.

Martín-Gutiérrez, J., Fabiani, P., Benesova, W., Meneses, M. D., & Mora, C. E. (2015). Augmented reality to promote collaborative and autonomous learning in higher education. *Computers in Human Behavior, 51*, 752–761. http://doi.org/10.1016/j.chb.2014.11.093.

Mulch, B. E. (2014). Library orientation transformation: From paper map to augmented reality. *Knowledge Quest, 42*(4), 50–53.

New Media Consortium. (2016). *NMC Horizon Report: 2016 higher education edition*. Retrieved from http://cdn.nmc.org/media/2016-nmc-horizon-report-he-EN.pdf.

CHAPTER 16*

# Virtual Reality Library Environments

*Jim Hahn*

## Introduction

What is virtual reality (VR), and how does it impact the research and teaching missions of the modern academic library? The *Horizon Report*, a resource that provides an annual accounting of technologies and trends relevant to higher education, has listed virtual reality as a technology that will likely be impacting higher education in the next two to three years (Johnson et al., 2016). The 2016 *Horizon Report*'s higher education edition defined VR as "computer-generated environments that simulate the physical presence of people and objects to generate realistic sensory experiences" (Johnson et al., 2016, p. 40). Related to the significance of technologies such as these, the *Horizon Report* authors note, "VR offer[s] compelling applications for higher education; these technologies are poised to impact learning by transporting students to any imaginable location across the known universe and transforming the delivery of knowledge and empowering students to engage in deep learning" (Johnson et al., 2016, p. 40).

However, while enthusiasm for virtual reality among educational technologists is high, the field of VR specifically for teaching and research applications is rather new, and so the educational application of VR is partly conceptual at this time. There are a variety of immersive games that are just becoming available on the consumer market. These early consumer products are illustrative of what will be possible with the new VR hardware released, or soon to be released, by several large technology corporations, including Sony, Facebook, and HTC. This chapter will review VR hardware, VR apps that are currently available, and hypothetical

---

* This work is licensed under a Creative Commons Attribution 4.0 License, CC BY (https://creativecommons.org/licenses/by/4.0/).

use cases for VR's application to research and teaching within library settings. The chapter will conclude with a summary of development requirements related to software design in VR, thoughts on future directions, and issues related to staying up-to-date with VR technologies in the future.

## General Virtual Reality Hardware

When technologists discuss VR hardware, there are several commonly agreed-upon technology solutions at play that incorporate a VR system. In virtual reality systems, the common technology solution is the headset. Headsets act to make the VR experience totally immersive. Unlike augmented reality (AR) applications, VR does not combine elements with the real world, a key exemplar of how VR is a truly immersive experience. Hardware requirements also include dedicated PCs with VR headsets, specific operating systems (OSs), and higher end graphics processing units (GPUs). These higher end graphic processors are required for processing the visual elements of an immersive world.

This chapter will highlight several notable new developments of compelling VR technologies in this area, with a specific focus on how these may be used or adapted to libraries. The specific VR consumer electronics reviewed in this chapter include Facebook's Oculus Rift headset (https://www.oculus.com/en-us), the HTC Vive (https://www.htcvive.com/us), and Sony's PlayStationVR3 (https://www.playstation.com/en-us/explore/playstation-vr). For a low-cost VR option and budget-constrained library IT department, a review of the mobile phone–based VR option Google Cardboard and its associated applications (https://www.google.com/get/cardboard/apps) is also provided.

## Contemporary Virtual Reality Hardware

The Oculus Rift (or as it is more commonly known, "the Rift") is one of the more popular contemporary VR systems to come to the awareness of library technologists. The Rift received much public acclaim and attention when news emerged that Facebook would acquire the Oculus company and brand (Solomon, 2014). Facebook made the acquisition in 2014 when a handful of contemporary VR headsets that comprise current consumer end products were still in the early stages of development. In 2014, VR headsets were mostly high-fidelity prototypes. However, as recently as the first quarter of 2016, consumer-ready VR products for the Rift were shipping to users. In order to use the Rift, the user must have a high-end computer with enough graphics processing available for rendering the virtual environment to the headset. To this end, the Rift website will also sell Rift-ready PCs. Common uses for the Oculus Rift are currently viewed as primarily gaming-based (Reisinger, 2016). However, with software development kits (SDKs), it is not out of the realm of possibility that other uses will come online shortly, a possibility that we'll explore further later in this chapter.

The HTC Vive is similar in nature to the Oculus Rift with regard to its configuration, since it requires a high-end graphics processor enhanced PC (Apple compatibility was not available at the time of this writing) in order to experience VR through the headset. Where the HTC Vive differentiates itself is that it allows a greater range of motion than is available from the Rift, which is generally considered a seated experience and not truly motion-enabled for the user. The HTC Vive, on the other hand, allows for a more active and motion-based experience. Reports indicate that users experience the feeling of truly walking through a virtual world and reaching out to touch objects (Eadicicco, 2016). By comparison, within the Oculus Rift, a game controller or a touch-enabled handheld device is required for touch experiences (Eadicicco, 2016).

One final product that experts in this field are following closely is the widely anticipated PlayStation VR. At the time of this writing, the PlayStation VR is not yet available to consumers and is anticipated to come to market in October 2016 (Weiss, 2016). The cost of the PlayStation VR is expected to be $399, which is a lower price point than the Oculus Rift ($599) and the HTC Vive ($799).

# Google Cardboard Virtual Reality Experience

Google Cardboard is a unique approach to VR experiences, which extends VR to a wider user base by leveraging existing mobile phone hardware at a fraction of the cost. The cardboard unit, or "viewer," is a product that can be ordered for as little as $20, and it simply uses a mobile phone to slide into the display case (https://www.google.com/get/cardboard/). Once users slide a mobile phone that is compatible with the Cardboard App into the viewer, they are able to hold the viewer to their eyes so that they are able to experience an immersive virtual reality scene, similar in nature to the VR headsets described above.

There is also a Google Cardboard SDK available for developing Cardboard apps (https://developers.google.com/cardboard/overview). Not every library will want to pursue this route, or if a library chooses to develop Google Cardboard apps, it may want to do so either as a part of a research grant; by outsourcing the coding to a developer group with previous Android development; or by or partnering with computer science research teams through which students may provide original and novel ideas toward library services. An alternative to the Cardboard experience is the Samsung Gear VR (http://www.samsung.com/global/

galaxy/wearables/gear-vr), which utilizes a Samsung phone and is built on similar technology to the Oculus Rift.

## Review of Virtual Reality Applications and Current Academic Uses

Google's Tilt Brush (http://www.tiltbrush.com) is a virtual reality app that may find uptake among artists, art students, and art researchers. At the time of this writing, the Tilt Brush app is available for the HTC Vive (http://www.fastcompany.com/3056668/googles-tilt-brush-is-the-first-great-vr-app).The demonstration of the Tilt Brush app shows a user totally immersed in making art, and the canvas is the virtual environment imposed all around. The user of this app therefore has a full range of motion to create. We may think of artists as being the natural users of this app, but there may be researchers in human development or ergonomics who wish to study user motion who may also find research and teaching uses with a compelling new application like the Tilt Brush.

The field of art history is seeing instruction-based VR use. A recent interview, entitled "The Promise of Virtual Reality in Higher Education," by Bryan Sinclair and Glenn Gunhouse, provides an accessible overview of several focused experiments in art history pedagogy with VR. Gunhouse has been working for a number of years to bring virtual worlds into art history teaching (http://www2.gsu.edu/~artwgg/atmos.htm). Among the projects he is bringing to art history teaching is the notion of teaching ancient Roman historical principles from "within" those objects, like a Roman house lecture taking part in a virtual world that recreates an ancient Roman house. Gunhouse notes that "what VR offers to my students is an increasingly true-to-life way of visiting places that we otherwise could not visit, either because they are very far away, or because they no longer exist" (Sinclair & Gunhouse, 2016). At this time Gunhouse says that the only thing holding back the full class from taking part in the VR-based lecture and learning is available equipment: "the lack of a classroom equipped with the necessary hardware" (Sincalir & Gunhouse, 2016).

Journalism is another discipline in which VR has found uptake by early adopters. As an example, there is a Cardboard app that the *New York Times* developed to tell stories in a more immersive way than would be possible with current mobile-based applications (http://www.nytimes.com/newsgraphics/2015/nytvr/). Being able to more fully tell a story is not solely relegated to journalism since digital presentations in many disciplines will rely on conveying information in a compelling way. Writing, communication, and rhetoric studies may also find uptake of the app by practitioners and students looking for new ways to engage their audiences.

The uses of VR for medicine are particularly intriguing for the ability to simulate complex human organs and the education of those who will be involved in surgical planning and surgical operation (Greenleaf, 1995). A variety of medical applications have already been reported on, from helping to treat anxieties like fear of flying to the development of treatments for PTSD (Carson, 2015). Uses in applied psychological studies have been reported, specifically in a case study where implications on the student of visual perception are explored (Wilson & Soranzo, 2015). One area within visual perception research made possible by VR headsets is new research examining visual illusions. Consider the types of visual illusions that perception researchers have made use of in their work, but applied in an immersive environment where the user is moving. Traditional studies of illusions on visual perception seemed to begin from the starting point that study participants were not moving or their movement was not an object of study—which is something the VR allows researchers to investigate (Wilson & Soranzo, 2015).

It is also the case that several STEM fields will have applications available soon for VR as well, due to these fields' needs for spatial and visual intelligence training. One recent report on utilizing virtual technology for engaging engineering focuses in the area of teaching mathematics and aerospace concepts to undergraduates (Aji & Khan, 2015). The authors of the work, "Virtual to Reality," describe using immersive experiences with flight simulation in order to engage teams of undergraduate students collaborating on test flight modules (Aji & Khan, 2015).

# Library Virtual Reality Use Cases for Research and Teaching

There are several virtual reality uses cases for research and teaching specifically within the library. In this section we review first-year-experience programming for VR, digital access to vendor content like e-books through VR, and data visualization and VR and conclude with thoughts related to makerspaces and VR production within libraries.

## *First-Year-Experience Programming*

The delivery of typical first-year-experience instruction has shifted over time as new technology has become available. What was once a tour of the building in person moved to online virtual tours as well as various introductory content about the library being available to students remotely. One of the first early exemplars on virtual worlds and library outreach was undertaken with experimental Second Life environments (Stimpson, 2009). Second Life is an online virtual world in which users can control avatars that can interact with other avatars and environmental objects. In these virtual environments, accessible from a browser, users navigated a multiuser world, including virtual library buildings and spaces that were designed by library staff. Though Second Life was not primarily designed for library outreach, practitioners were engaged in creating virtual library outreach where users could come to experience in Second Life another "branch" of the library. These branches were not heavily used, and not all libraries made efforts to develop programming specifically targeted at this VR branch (Stimpson, 2009). More recently, Second Life developers have worked to use Oculus Rift to explore Second Life (Linden Lab, 2014). Whether the new Oculus Rift–based gateway into Second Life will convince more users and library practitioners to try the virtual world remains to be seen.

Self-guided podcast and iPod app tours have also been utilized as a novel technology to provide tours and introductions in libraries (Mikelle & Davidson, 2011). This provides an example of the historical continuities among older technologies and new, emerging consumer products that have been able to provide new or more efficient services. Library technologists may be interested in developing immersive virtual library and virtual campus tours that take place beyond the campus. If they do so, students who are excited about visiting campus and are not nearby can plan a virtual reality visit. This can help drive down travel costs for students so that they are required to make the trip to campus only when they begin their studies. Partnerships with admissions departments and other new-student programs are advisable, since these groups will be stakeholders in any technology application that helps recruit students to campus. Pooling resources can help to defray development costs and help to promote innovative design ideas.

## *HoloBooks: E-book Reading in Virtual Reality Environments*

The challenges facing academic libraries to continue to steward print resources while investing heavily in online content has led many in the public sphere, as well as from the academic library profession, to question the need and use of legacy print collections. Lee, in the classic work "What Is a Collection?" evaluates this intermediary state of collection development and argues compellingly for redefining collections in libraries as information contexts (2000). Researchers at the University of Illinois library have begun crafting a project to develop virtual reality e-book reading experiences allowing library users to explore and read e-books from virtual worlds. The proof-of-concept system was initially targeted at the Microsoft HoloLens, a mixed-reality technology. The Microsoft HoloLens does not qualify as a truly immersive VR experience, since it uses targets in the real world along with a headset to create projections of digital surrogates onto real-world items.

The premise of the HoloBook project is to develop functionality to generate a page on a physical paper-like target to mimic reading a printed item. HoloBooks could apply to several or all of the VR headsets described in this chapter. Rather than using a paper-like target in the case of the Microsoft HoloLens, the completely immersive VR environment would handle all of the rendering and display of a HoloBook.

Development of HoloBook features will include emulating page turns, annotations, and highlighting functionality. By developing the HoloBook reading experience, researchers are interested in testing the hypothesis that reading digital items from print-like surrogates can support increased comprehension of text. Of relevance to the research and teaching needs that we see in academic libraries, consider the visualization of electronic resources, which become in a sense more tangible for the students who use only stacks-based browsing to locate items of relevance to their research. Librarians, educators, and publishers are faced with the challenge of providing access to digital content while still maintaining legacy print collections. The HoloBooks experiment will address access and use problems inherent in digital library collections of the networked era, including their highly disparate nature (many vended platforms serving licensed content) and their increasing intangibility (the move to massively electronic or e-only access in libraries and information centers). By developing a HoloBook reading experience, researchers will operationalize the transformation of digital-only content into print-like experiences and augment this mixed-reality experience with digital VR research support. Studies have shown that in some cases, reading comprehension may be lower on e-readers when compared to reading of nonfiction in print (Mangen & Kuiken, 2014). The proposed HoloBook project aims to provide contextual support for comprehension and learning. Researchers hypothesize that with VR reading, digital text will achieve a comprehension equivalent to print-based reading. By developing a unique HoloBook reading experience, librarians can provide integrated research support at the point of need. The current traditional reading experience does not allow opportunities for the integration of digital resources and research support—new HoloBook features and value-added services can support undergraduate students, especially as they go about completing research papers in their critical first years of study. HoloBook research support is responsive to students' contextual needs—since undergraduate student research in the digital era suffers from inadequate guideposts for knowing where to start, and students are seeking context when they begin research. This context is increasingly difficult to obtain in the online sphere (Head & Isenberg, 2009).

## *Statistical Visualization in Virtual Reality*

Consider a VR use case that responds to the needs for data visualization and analytics within libraries. VR could help support visualizing assessment data. In academic libraries in particular, the need to support data analysis is great. It is not uncommon for researcher data to be derived from several different places—or in the data science terminology, data exists in silos within research organizations. Virtual exploration of data may help support the visualization from multiple disparate data sources and help decision makers explore data while at the same time help them to understand where gaps in their data-based decision making exist. Visualizing data is of course only one small component of data analysis—but it is a sector of the data enterprise that virtual reality could make more efficient, effective, and even more dynamic and enjoyable to explore. As an example from commerce, the start-up company CodeScience is at work developing an Oculus Rift VR app to visualize data from Salesforce (Rima, 2015). The Salesforce platform is a well-known customer relationship management tool. Therefore, a company with a Salesforce data API may likely be interested in quickly generating a picture of open leads, contracts, or client requests that may need attention.

Another example of the importance of VR for data visualization includes addressing the visualization of data with many dimensions. According to Donalek and colleagues, "The more dimensions we can visualize effectively, the higher are our chances of recognizing potentially interesting patterns, correlations, or outliers" (2014, p. 610). Using VR for data visualization would be particularly helpful in efforts to provide data curation where very large sets of data are curated by library professionals in collaboration with scientists. Researchers from Caltech also evaluated several tools for VR-based data visualization in the sciences and found that by employing VR headsets and common virtual world engines (like Second Life) that "these technologies give us a significant, cost-effective leveraging: their rapid development is paid for by the video gaming and other entertainment industries, and they are steadily becoming more powerful, more ubiquitous, and more affordable. They offer us an opportunity where any scientist can, with a minimal or no cost, have visual data exploration capabilities that are now provided by multi-million dollar cave-type installations, and with a portability of their laptops. Moreover, they open potentially novel ways of scientific interaction and collaboration" (Donalek et al. 2014, p. 613).

The notion of storytelling surfaces several times when discussing the capability of virtual reality. Telling stories with data would be a valuable use case for VR headsets. Consider the immersive possibilities of engaging with stakeholders by way of immersive data representations and findings. A compelling analysis of assessment and other types of learning data is an area ripe for innovation and piloting.

## *Makerspaces and Virtual Reality*

Other considerations within libraries include integrating experimentation with VR experiences into makerspaces in libraries. Media creation and media development would find nice dovetails into virtual reality development. Students and faculty creating media may be interested in gaining experience in immersive storytelling.

One strand that deserves considering is the makerspace movement and its applicability to VR development, experimentation, and exposure to VR capabilities. According to the *NMC Horizon Report*, "Makerspaces are informal workshop environments located in community facilities or education institutions where people gather to create prototypes or products in a collaborative, do-it-yourself setting" (Johnson et al., 2016, p. 42). One can think of makerspaces as being areas by and for content creators—they are inhabited by people who are actively designing and crafting content or otherwise producing a tangible output of work. Makerspaces offer the promise to move academic institutions from places that primarily consume to places that are capable of design and production. These are the skills needed in the twenty-first-century workplace.

How will creative makers choose to engage in this medium? Creating and crafting video will be a part of this, to be sure. In the more traditional or classical conception of a makerspace, we might think of the space as being an area for video creation hobbyists. Extending from this video hobby, we might theorize that those with an interest in video creation and the 360 degree affordances that the VR headsets allow may be interested in the Jump Toolkit, which is intended to create 360 degree video viewed from within Google Cardboard (http://www.google.com/get/cardboard/jump/).

There are gaming development elements to VR as well, of course, those that blur the lines between narrative storytelling, illustration, and animation. It may certainly be the case that illustrators or those doing graphic design in makerspaces may help provide like-minded individuals with the space and materials for prototyping portions of VR experiences.

## Developer Resources

Each of the VR headsets discussed above has a unique developer environment. As we previously noted while introducing the Google Cardboard SDK, software development can be a resource-intensive operation, which is also the case for virtual reality development. The following are SDKs and services to use when developing for each of the headsets described in this chapter:

- Oculus SDK is, at the time of this writing, in version 1.3. There is a helpful developer blog available at http://developer.oculus.com that periodically reports specific upgrades to the SDK and other developer-friendly assistance in creating applications for Oculus hardware.
- Due to the relative newness of the HTC Vive, its developer portal is just getting started with documentation and resources. There is the beginning of a portal here: https://www.htcvive.com/us/develop_portal, and an interesting open VR GitHub portal with some vendor-agnostic tools here: https://github.com/ValveSoftware/openvr/wiki/API-Documentation.

- Sony's PlayStation VR similarly does not yet have too much documentation on development resources for it. The Unity engine may be used in a variety of virtual reality systems and may be used for PlayStation VR development as well. Helpful developer documentation for the Unity engine is available here: https://unity3d.com/unity/multiplatform/vr-ar.

Sinclair and Gunhouse note that development work requires "a high-end desktop computer with fast processors and a good CPU. The software required includes a 3D modeler of some sort" (Sinclair & Gunhouse, 2016).

Compelling academic partnerships for VR development may exist for developing immersive experiences. Recent work in partnering with computer science faculty and students has led to innovative services and products that libraries are using in operations (Hahn, 2015). When undertaking these partnerships, there are some general themes to be aware of if engaging with undergraduate student coursework. These include being mindful of time constraints with student scheduling. Unlike full-time library developers, students will not be able to put in full days of work on an experiment project. However, they can contribute pieces or modules of functionality toward software development goals. So, while students are not professional developers, their prespecialization does come with certain advantages to a library makerspace for VR. The advantages to being prespecialized sometimes mean that the student does not understand the limitation of the domain—which can actually work to the advantage of those developing ideas; idea generation for virtual reality should incorporate student preferences and expectations for these environments. In this way, service uses can be built in early in the creation process of VR so that library-specific application is going to be useful for students and also respond to identified use cases.

## Future Directions

This chapter is intended to bring the reader up to speed with several new areas of virtual reality and its applicability for teaching and learning in library settings. In order to stay up-to-date with VR technologies in the future, consider following these media outlets:
- *Wired* magazine is a long-running popular science and technology magazine that regularly reports on general-interest technology. Its focus on consumer electronics and popular culture makes it highly readable and relevant to up-and-coming trends within virtual reality in general. It is not library-specific, but it will help library leaders understand and track newer options in virtual reality as they become more generally available.
- The Verge (http://www.theverge.com) is a newer entrant (founded in 2011) in the news-tech field. It does provide several long-form reads

that treat technology news in depth, and provide additional deep understanding for newer consumer-facing products, similar to *Wired* magazine.
- The *New York Times* is increasingly covering cutting-edge technologies. As this chapter was being crafted, a helpful review of the Oculus was published (Chen, 2016).
- EWeek (http://www.eweek.com) is a shorter form roundup of quick news items. Its social media pages will help provide a snapshot of current virtual reality and other upcoming technologies. Usually, its posts include business-focused content—which can be useful for organizationally driven libraries in academic settings.

Each of the above-named technology organizations can be followed from its social media account, and their Twitter accounts are particularly useful to follow for timely updates on the continuing development of virtual reality. To dive deeper into academic-based research by computer scientists in mixed reality, consider following the ISMAR symposium (http://www.ismar.net). The international ISMAR symposiums have been the location for cutting-edge research in this domain since 2002.

Virtual reality consumer products offer new functionality for libraries, but they are also a largely untested technology. However, since the current wave of consumer VR products represents a profound advancement over previous iterations of head-mounted displays and previous virtual worlds, the trend should not be overlooked. Rather, librarians and academics may find that VR options can make possible teaching and research services that were previously unavailable, increasing students' engagement in libraries and higher education alike.

# References

Aji, C. A., & Khan, M. J. (2015). Virtual to reality: Teaching mathematics and aerospace concepts to undergraduates using unmanned aerial systems and flight simulation software. *Journal of College Teaching & Learning, 12*(3), 177–188. http://dx.doi.org/10.19030/tlc.v12i3.9342.

Carson, E. (2015). 10 ways virtual reality is revolutionizing medicine and healthcare. *TechRepublic*. Retrieved from http://www.techrepublic.com/article/10-ways-virtual-reality-is-revolutionizing-medicine-and-healthcare/.

Chen, B. X. (2016, March 28). Oculus Rift Review: A Clunky Portal to a Promising Virtual Reality. *New York Times*. Retrieved from http://www.nytimes.com/2016/03/31/technology/personaltech/oculus-rift-virtual-reality-review.html?smid=tw-share.

Donalek, C., Djorgovski, S. G., Cioc, A., Wang, A., Zhang, J., Lawler, E., Yeh, S., Mahabal, A., Graham, M., Drake, A., Davidoff, S., Norris, J., & Longo, G. (2014). Immersive and collaborative data visualization using virtual reality platforms. In *Big Data (Big Data), 2014 IEEE International Conference on* (609–614). IEEE. Retrieved from http://arxiv.org/ftp/arxiv/papers/1410/1410.7670.pdf.

Eadicicco, L. (2016). Virtual reality buyer's guide: Oculus Rift vs. HTC Vive vs. Samsung Gear VR. *Time.Com*, 4.

Greenleaf, W. J. (1995). Virtual reality applications in medicine, *WESCON/'95. Conference record*. 'Microelectronics Communications Technology Producing Quality Products Mobile and Portable Power Emerging Technologies', San Francisco, CA, USA, 1995, pp. 691–696. doi:10.1109/WESCON.1995.485484. Retrieved from http://ieeexplore.ieee.org/stamp/stamp.jsp?tp=&arnumber=485484&isnumber=10332.

Hahn, J. (2015). The student/library computer science collaborative. *portal: Libraries and the Academy, 1*(2), 287–298.

Head, A., & Eisenberg, M. (2009). Finding Context: What Today's College Student Say about Conducting Research in the Digital Age. *Project Information Literacy Progress Report, University of Washington's Information School*. Retrieved from http://projectinfolit.org/images/pdfs/pil_progressreport_2_2009.pdf.

Johnson, L., Adams Becker, S., Cummins, M., Estrada, V., Freeman, A., & Hall, C. (2016). *NMC Horizon Report: 2016 Higher Education Edition*. Austin, Texas: The New Media Consortium.

Lee, H-L. (2000). What is a collection? Journal of the American Society for Information Science, 51(12), 1106–1113.

Linden Lab. (2014, October 13). Oculus Rift DK2 Project Viewer Now Available. *Second Life Blogs: Feature News*. Retrieved from https://community.secondlife.com/t5/Featured-News/Oculus-Rift-DK2-Project-Viewer-Now-Available/ba-p/2843450.

Mangen, A., & Kuiken, D. (2014). Lost in an iPad: Narrative engagement on paper and tablet. *Scientific Study of Literature, 4*(2), 150–177. http://doi.org/10.1075/ssol.4.2.02man.

Mikkelsen, S., & Davidson, S. (2011). Inside the iPod, outside the classroom. *Reference Services Review, 39*(1) 66–80.

Reisinger, D. (2016). Should you buy an Oculus Rift vr headset: 10 factors to consider. *Eweek*, March 30, 7.

Rima, Gianna. (2015, October 9). *Using Oculus Rift and Virtual Reality to Visualize Data on Salesforce* [Video file]. Retrieved from https://youtu.be/oboOkHmwr_Q.

Sinclair, B., & Gunhouse, G. (2016). The promise of virtual reality in higher education. *Educause Review*, March 7. Retrieved from http://er.educause.edu/articles/2016/3/the-promise-of-virtual-reality-in-higher-education.

Solomon, B. (2014). Facebook buys Oculus, virtual reality gaming startup, for $2 billion. *Forbes.Com, March 25*, 1.

Stimpson, J. D. (2009). Public libraries in Second Life. *Library Technology Reports, 45*(2), 13–20.

Weiss, T. R. (2016). Sony unveils a $399 PlayStation vr headset. *Eweek*, March 16, 2016, 1.

Wilson, C. J., & Soranzo, A. (2015). The use of virtual reality in psychology: A case study in visual perception. *Computational & Mathematical Methods in Medicine, 2015*, 1–7. doi:10.1155/2015/151702.

# Works Consulted

The rise and fall and rise of virtual reality, http://www.theverge.com/a/virtual-reality.

Is 2016 the year virtual reality finally goes mainstream?, http://www.npr.org/sections/alltechconsidered/2015/12/30/461432244/is-2016-the-year-virtual-reality-finally-goes-mainstream.

Virtual reality, https://en.wikipedia.org/wiki/Virtual_reality.

Google Cardboard, https://www.google.com/get/cardboard/.

CHAPTER 17*

# Wearable Technologies in Academic Libraries
## Fact, Fiction and the Future

*Ayyoub Ajmi and Michael J. Robak*

## Introduction

Nick Moline, a developer and early Google Glass Explorer, can still recall Google's mantra when he was first introduced to the wearable device: "If you can bring technology closer to you, you can actually get it out of the way" (Moline, personal communication, December 29, 2015). Similarly, Steve Mann, a researcher and inventor widely known as the father of wearable computing once wrote that "miniaturization of components has enabled systems that are wearable and nearly invisible, so that individuals can move about and interact freely, supported by their personal information domain" (Nichol, 2015). Today's wearable devices are the continuation and evolution of decades of research and development. This transition began with devices designed to be worn as backpacks, such as the 6502 multimedia computer designed by Steve Mann in 1981, evolved to a one-handed keyboard and mouse connected to a head-mounted display produced in 1993, and then advanced further into a wrist computer made available the next year. The first commercially available wearable device, however, was the Trekker, a 120 MHz Pentium computer with support for speech and a head-mounted display, which sold for $10,000 (Sultan, 2015). These early wearable devices, however, were characterized by limited functionality and bulky design. By the mid 2010s, fitness tracker devices emerged with

---

* This work is licensed under a Creative Commons Attribution-ShareAlike 4.0 License, CC BY-SA (https://creativecommons.org/licenses/by-sa/4.0/).

their attractive designs targeting sport and fitness enthusiasts. More recent fitness trackers blend smartwatches with multiple other functionalities, combining health and activity monitoring as well as networking capabilities.

There are many factors that contributed to the rapid proliferation of wearable devices in the last five years. These factors include the advent of more reliable Internet access; the ubiquity of smartphones; decline in cost of sensors, cameras, and processing power; and finally, a flourishing app ecosystem (Mind Commerce, 2014).

## Market Analysis

Predicting growth rate in the wearable devices market is difficult, if not impossible. Many of the devices available today are still in prototype format, and many are starting from scratch with a high probability of increasing functionalities as new technological advances occur. According to Gartner's (2015) *Hype Cycle of Emerging Technologies*—an annual report representing the maturity, adoption, and social application of specific technologies—wearable devices are still five to ten years away from their mass adoption. In 2015, they leave the peak of inflated expectations and now slide into the trough of the disillusionment cycle, where interest decreases as experiments and early implementations fail to deliver. While few companies will take action, only those that demonstrate innovative solutions capable of satisfying early adopters needs will succeed in reaching the plateau of productivity (figure 17.1) (Gartner, 2015).

**FIGURE 17.1**
Hype Cycle for Emerging Technologies 2015

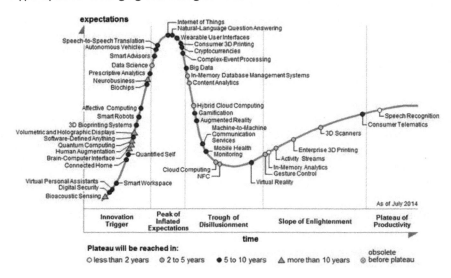

According to Forrester's 2015 consumers and technology report, 21 percent of US adults online use a wearable device (Fleming, 2015). This adoption, concentrated in activity-tracking devices and smartwatches, is mostly driven by the mass adoption of smartphones and the proliferation of applications that created a social engagement that was not available just five years ago.

Activity-tracking devices, led by Fitbit, provide easily tracked fitness-related metrics such as steps walked, distance walked or run, sleep and activity time, and other valuable information designed to promote a healthy lifestyle. However, as new products hit the market, the volume of specialized fitness and health-care wearables is expected to shrink from 60 percent of the wearable market in 2014 to 10 percent in 2018, when multifunction consumer wearable products will dominate (Mind Commerce, 2014). Mind Commerce (2014) also projects that the volume of wearable computing devices will grow from under one million units in 2014 to 178 million in 2019, dominated by smart glasses and smartwatches.

Computer technology advanced from mainframes to desktops, then laptops and palmtops, and is now moving onto, and into, the human body by way of wearable computers (Rainie, 2016). This type of gadget provides the ultimate in network access—hands-free, heads-up operation with complete mobility. Among the factors that contributed to the current surge in wearables are the advent of the Internet and broadband connection, proliferation of smartphones and other mobile devices, rapid growth in sensors and other micro-electromechanical systems, and an increasing availability of apps. The success of wearable devices will, in fact, depend tremendously on third-party developers who can integrate wearable devices and their features into existing or purpose-built experiences.

## Google Glass Explorer

Google Glass is, without a doubt, the device that brought the discussion of wearable technologies to the public with its innovative design and features, as well as numerous controversies. Google Glass is a head-mounted display that presents data in the wearers' field of vision without the need to look away from their normal viewpoints (Google Glass, 2016). The device is tethered to a smartphone, and most of its functions are voice-activated. Google X, Google's research and development facility that is responsible for the Google Glass and Google driverless cars projects, among others, succeeded in packing a battery, a display, a camera, and all the processing power needed to run the device into a compact 36-gram frame (figure 17.2).

### FIGURE 17.2
Google Glass technical characteristics.

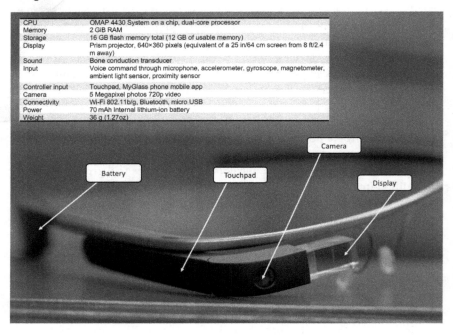

Google Glass was first demonstrated to the public during the 2012 annual software developer conference (Google I/O). The Explorer Edition was then made available to Google developers only; later, it became available for purchase for selected users during the #ifihadglass Google+ and Twitter contest, for a purchase price of $1,500. From April 15, 2014, to January 19, 2015, the prototype device was available for sale before Google announced the end of its Open Beta Google Explorer Program (Google, 2015).

At the University of Missouri–Kansas City's Leon E. Bloch Law Library, the authors joined the Google Explorer Program in April 2014 with the goal of testing and exploring the potential of the device in academic libraries. Before entering the Open Beta program, the authors brainstormed a few use cases in which they believed the device could improve productivity and collaboration. Due to the lack of applications dedicated to education or that can serve an obvious purpose in a library environment, the use cases were solely focused on the first-person perspective experience that the device provides using its built-in camera. When the law library joined the program, Google already had many promotional videos demonstrating Google Glass, among them, a virtual field trip to the Large Hadron Collider of CERN (the European Organization for Nuclear Research) using

Google Glass (Google Glass, 2013). The video absolutely inspired the authors to immediately consider using the device to offer virtual tours for potential students and other live fieldtrips. However, the video feature was later dropped during the XE16 firmware update. Nevertheless, the library produced a hands-free video tour of the library using Google Glass, but not without a few workarounds.

The Explorer Edition of Google Glass is set to record ten seconds of video as a default. A user can manually extend the duration of the video indefinitely, or as long as there is battery power left, which is a maximum of forty-five minutes of video recording in a single full charge. However, the limitations of the videos produced through Google Glass are not related only to the short battery life, but also to other noticeable drawbacks in the software and hardware. Rice University published a study of Glass's power and thermal characteristics in which it found that the device can easily reach 50 degrees Celsius when using power-hungry applications such as video and GPS navigation (LiKamwa, Wang, Carroll, Xiaozhu Lin, & Zhong, 2014), making the wearer unlikely to record video for an extended period of time. Other issues encountered when creating library video tours were the poor quality of the picture when the device is used in low-lighting situations and the video instability (think bobblehead dolls) when the wearer is moving. In a different use case, students used Google Glass to record their interviews with potential clients in a role-playing exercise. Students expressed their satisfaction with the device as its hands-free feature enabled them to focus more on their assignments rather than being distracted with the technology. Despite a limited interaction with Google Glass, the authors remain confident that, paired with the right applications, it can offer benefits for both libraries and their users. The technical limitations reported in the early versions of Google Glass can eventually be fixed over time by Google.

Among the first Explorers who participated in the Google+ and Twitter #ifihadglass contest, many educators and librarians shared how they could use Google Glass to improve the learning experience inside and outside the classroom. Google also maintained an online discussion forum for Google Glass Explorers where ideas were shared and many connections were made. Adam Winkle, a K–6 science teacher from Florida who received the first edition of Google Glass with the help of an after-school program grant, has been very active within the Explorer Community. Winkle used Google Glass on a daily basis and also used it to produce science, technology, engineering, and mathematics (STEM) videos for his students. His experience was very positive as it helped him to create educational material quickly and efficiently without affecting his teaching ability and the time he spends helping students in class (Winkle, 2015). In addition to exploring Google Glass at a personal level, Winkle also works closely with EduGlasses, a start-up specialized in developing educational applications and services for administrators, teachers, and students through the use of smart glasses such as Google Glass and Epson Moverio (Winkle, 2015).

While developing custom applications to solve real-life problems is the best way to take advantage of Google Glass features, not every Explorer has the knowledge or technical capability do so. Many of the early Google Glass Explorers were limited to the few applications that Google made available through its app store. Others were able to side-load third-party applications, such as those of EduGlasses, which requires some expertise and risks voiding the warranty on the device. Jenn Waller is one of the early adopters who used the device as-is, out of the box. Waller, a librarian at Miami University in Ohio, enjoys demonstrating Google Glass to her students and colleagues. She acquired the first edition of Google Glass with the help of her library's innovation grant, and she has used it in her library instruction classes and individual office appointments to introduce students to wearable technologies and to discuss concepts of privacy and sharing in the digital age. Waller also helped her library purchase additional devices through a technology grant. The new devices have been added to the circulation collection, and students are encouraged to check them out and explore ways they can use the devices or develop new apps for them. However, Waller didn't have a great experience using Google Glass personally, as many hardware and software obstacles prevented her from fully utilizing its features. Nevertheless, she believes that, as a librarian, her role is to help users find information in whatever medium or format fits their needs (Waller, 2016).

Roxann Riskin, on the other hand, a library technology specialist at Fairfield University, embraced Google Glass in both personal and professional contexts in many creative ways. Riskin, who acquired the first edition of Google Glass with her personal funds, finds the device very intuitive and easy to use. She uses Glass on a daily basis for communication and collaboration, as well as sharing its potential with students and colleagues. Her interest in Glass opened the door to other creative collaborations. With her project partner Rick Sare, who is a professional truck driver and also an early Google Glass Explorer, they published four educational books for children using unedited photos captured through Glass while collaborating through Google Cloud platforms. Riskin also used her Google Glass to capture photographs for her personal poetry book *Glass on the Beach* (Riskin, 2016).

Google Glass Explorer was Google's Open Beta program, which offered the device to consumers and individual developers. In two years, Google amassed an incredible amount of feedback from thousands of users testing the device in real-life situations. When Google ended the program in January 2015, it also ended its collaboration platform, support, and hardware and software releases for the consumer version of the device. Google stated that the end of the Explorer Program is a graduation of Google Glass from a proof of concept to real product and that the team will take what they learned from the early adopters to focus on an enterprise version of the device (Google, 2015). The focus has shifted now to Glass at Work and what smart glasses can bring to the workplace through its Glass Certified Partners. These partners have direct access to Google's technical support and unlimited inventory of Google Glass devices. Among the companies Google has partnered with for its Glass at Work initiative are Pristine, the creator of EyeSight, a secure two-way video communication platform for Glass dedicated to health care; GuidiGO, the creator of virtual tour guides for museums and cultural institutions; and the American Medical Association, specialized in medical field solutions in telemedicine. While the partners are working on new innovative solutions that will bring unique experiences to Glass users, Google is also working on a new iteration of its hardware in order to overcome the issues reported by users of its first edition, such as weak connection, small prism, and short battery life.

In December 2015, Google filed an application for equipment authorization with the Federal Communications Commission detailing what appears to be an Enterprise Edition of Google Glass (OET Exhibits List, 2016). The photos attached to the filing shows a foldable device with a larger prism (see figure 17.3). Other upgrades are expected to improve the connectivity of the device to better support video transmission as well as efficient processing power to improve its battery life. After all, if the device is to be used in fast-moving and rough industrial or emergency situations, short battery life or weak signal won't be tolerated.

**FIGURE 17.3**
View of a foldable version of Google Glass.

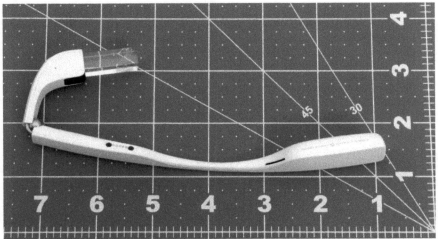

With the emergence of Google Glass, Google deserves the credit for bringing smart glasses technology, and wearables in general, to the consumer market. However, every player in consumer electronics is working on some version of a wearable device. Other manufactures are also taking advantage of this increased awareness and working hard to secure a share in the market.

## Other Smart Glasses

Sony released its own developer edition of smart glasses called SmartEyeglass featuring holographic binocular eyewear capable of providing an augmented reality experience by superimposing text and images onto the users' natural field of view. While this device has similar functionalities to Google Glass in term of connectivity, its camera, and its plethora of sensors, it also includes a detached controller/battery pack and currently works with Android-based smartphones only (SmartEyeglass, n.d.). At the 2015 Consumer Electronics Show (CES), Sony revealed a concept model of a single-lens display that can be mounted to existing eyewear under the working title of SmartEyeglass Attach! While the modular version is still in proof of concept, it is currently available in the market for developers and interested users (Sony, 2015).

Epson is another manufacturer working on wearable technologies coming in a variety of shapes and formats. Its smart glasses line of products Moverio BT-100, BT-200, and BT-300 are similar to Sony's smart glasses in term of features and hardware, with a holographic dual display and external controller. Epson also released Moverio BT-2000, a smart headset with a robust design and advanced features optimized for industrial applications (Epson, n.d.).

Vuzix, a developer and manufacturer of smart glasses and video eyewear products, released the first commercially available smart glasses, Vuzix M100. The M100 features a monocular display and functionalities similar to Google Glass. Vuzix is also working on other types of smart glasses targeting enterprise and consumers with a newly updated version of its monocular display and new augmented reality smart glasses, AR3000 (Vuzix, n.d.).

Other big names are also working on their own versions of smart glasses. Amazon, Apple, and Baidu, China's most used search engine, have all received patents for devices that enable users to access information or entertainment directly in their line of sight. Amazon was granted a patent for a special smart glasses that enable users to stream content from other devices right through the lens of the glasses (Kim, 2015). Apple's iGlass regulates the use of LCD displays of the glasses to cover the user's direct and peripheral vision (Mind Commerce, 2014). Baidu's prototype device, known as Baidu Eye, will leverage the company's strengths in image search and facial recognition (Lee, 2013).

## Activity Trackers

While manufacturers are still experimenting and trying to figure out what functions and design elements will improve smart glasses and increase their adoption, activity trackers have already reached a large set of the population with their basic functionality and appeal to niche audiences interested in health and exercise.

These devices, generally worn on the wrist, track wellness and sync wirelessly with a phone to send data about step counts, calories burned, distance, pace, activity time, sleep patterns, and other vital information from the wearer. Among the most popular activity trackers on the market today are the Jawbone UP band, Fitbit Flex band, and Nike+ FuelBand. However, with Google's and Apple's introduction of smartwatches, which offer, among other features, similar fitness tracking functionality, it is becoming difficult to distinguish between fitness-specific devices and smartwatches

## Smartwatches

Many industry players have ventured into the market of smartwatches in the past; however, due to technological constraints at the time, they failed to capture the imagination and interest of the mainstream consumer. Today's smartwatches are seen as an extension of smartphones, displaying notifications, calls, calendars, news, and other smartphone-like applications, in addition to displaying time. Moreover, analysts argue that many smartwatch manufacturers simply attempt to replicate smartphone functions on a miniaturized screen, without taking into consideration that consumers wear watches as accessories and fashion, which could mean that manufacturers may secure a niche market for a short time, but that it will eventually lead to both user frustration and an inferior user experience (Mind Commerce, 2014). The proliferation of smartphones today and the dominance of Apple and Google in this sector, combined making up 90 percent of the mobile market, it is likely that smartwatches will also be dominated by these two players. Apple released its smartwatch in spring 2015, targeting the luxury end of the market. With its existing large customer base, the Apple Watch easily secured a leading spot in the market. On other hand, Google released its own operating system, Android Wear, designed for smartwatches to complement smartphones running on the Android platform. Google partnered with several hardware manufacturing companies to release an armada of smartwatches unique in their design and function, targeting all sorts of applications and user groups.

Pebble is another breed of smartwatches competing with Apple and Google. This smartwatch is compatible with both Apple and Android-based smartphones and features an e-paper display technology, providing a longer battery life, an always-on feature, and a lower price tag than its competitors (Wikipedia, 2016).

## Other Wearable Devices

So far, this chapter has focused on consumer wearable technologies with immediate and direct impact on users, such as smart glasses, smartwatches, and activity-tracking devices. However, there are industries where wearables are making a considerable impact, and many industry players are competing to secure their place in niche markets such as gaming, health care, and the military.

From smart glasses to virtual reality headsets and smart clothing to simulate virtual environments for players, wearables for gaming are available in many shapes and formats. Gaming is one of the largest drivers of innovation in the wearable market, with a potential growth of $1.5 billion in sales revenue by 2019 according to Mind Commerce's mid-range forecast for gaming devices (Mind Commerce, 2014). In recent years, the health-care market has seen an increase in the adoption of wearable devices targeting the monitoring of physiological and vital signals, on-site personal patient care, and remote health care, all encouraged by recent advances in sensors, connection, and other technologies. Wearable health-related devices are expected to produce groundbreaking innovations that will help patients manage their chronic diseases as well as prevent medical complications. Wearables in the military have the same characteristics as the ones available on the consumer market. However, the military devices are built to operate under harsh conditions and provide the wearers with enhanced tactical awareness and advantages on the battlefield.

## Wearables in Academic Libraries

Libraries in general, and academic libraries in particular, play a role in introducing their communities to a variety of new technologies that they otherwise can't afford or know little about. In the absence of clearly understanding how these new technologies can be used in libraries, due to their novelty, in many cases librarians themselves are learning as they use the technologies. Tom Bruno from Yale University libraries introduced Google Glass to his community and sat back to see what they would make of the technology (Bruno, 2015). He introduced a program during the summer break in which members of the Yale community submitted project ideas. These projects were reviewed, and based on the practicality, technological feasibility, and merit of their submitted projects, the community members were able to check out Google Glass devices from the library (Bruno, 2015). The novelty of the technology, combined with the curiosity of students, allowed for unique opportunities to introduce other library services and information research concepts otherwise difficult to promote. Waller (2016), for example, used her Google Glass to introduce students to copyright challenges surrounding wearables and mobile devices in general.

Many academic institutions are also experimenting with wearables as instructional devices. The University of California, Irvine School of Medicine collaborated with Prestine, one of the Google Glass for Business partners, to launch a pilot program using the wearable device in anatomy courses and clinical skills training (Irvine, 2014). Students and instructors were expected to take advantage of the hands-free and voice command feature to access patients' information and record students' activities. At the authors' institution, the University of Missouri–Kansas City School of Law, students used Google Glass for similar purposes to record their health law course assignment. The students were required to select a friend, classmate, or family member to act as a simulated client for whom they were to provide end-of-life decision-making documents. The first person perspective of the video recording helped students improve their communication skills by allowing them to see their interviewees' reactions, which are otherwise very difficult to capture using regular cameras. Smart glasses have found a particular niche market in visual and demonstration-based education. The potential of bringing live experiments and demonstrations to learners in real time, or on demand, can positively impact the learning outcome. In a unique experiment, a group of surgeons from Queen Mary University of London's Medical School used Google Glass to record and live stream a surgical teaching session to more than 13,000 health-care professionals and members of the public from around 115 countries (Sultan, 2015). As wearable technologies become more reliable and reach mass adoption, academic libraries will again be asked to help bridge the gap between the haves and the have-nots in the same manner they are currently providing access to laptops, e-readers, digital cameras, and all sorts of devices through technology "petting zoos" and instructional courses.

Wearable devices can also help to improve staff productivity in libraries. Compared to mobile scanning units popular among libraries, smart glasses such as Google Glass can represent a cheaper alternative. Bruno (2015) argues that staff members can create new workflows for library work. For example, book returns, and other library processes requiring barcode scanning can potentially be expedited because the wearable can serve as the check-in scanner and require fewer steps in the shelving process. Other applications can also find their place in a library setting, such as Word Lens, an app that provides real-time translation of text captured through smart glasses that can potentially be used to help identify foreign language material or help break the language barrier with library users.

Perhaps the biggest potential of wearable devices in libraries is the ability to engage users by creating content suitable for these new mediums. Users can immerse themselves in library collections through smart glasses by simply glancing at a book cover. Abstracts, reviews, video excerpts, and all types of related content harvested from multiple sources can be displayed in the user's line of sight. Wearable devices, when combined with location-based information and services such as Apple's iBeacon technology, QR codes, or near-field communication (NFC) tags can allow for the broadcasting of targeted messages and information to users in locations where Wi-Fi and cell phone signals are nonexistent. Libraries have been using these services to push notifications to users, who opted into the service, about events, new books, items due, and other highly customizable messages. While research has been centered around wearable devices from the user's perspective, wearables can also come in the form of wireless sensors that can be embedded in the library's physical space, allowing for monitoring of vital information such as occupancy, noise level, temperature, usage, foot traffic, and traffic flow. Data logged can help in decision making and can also be shared back with users through their wearables in a readable and aggregated format.

## Challenges

Wearable devices are still in their early stages of development and adoption. While health tracker devices have shown a great potential in the fitness and health-care fields, smart glasses, smartwatches, and other wearables are still missing a killer application that can boost their adoption in the same way that iTunes helped drive iPod sales (Mind Commerce, 2014). Early adopters of wearables in libraries are then limited to out-of-the-box applications, and the use cases presented mostly remain hypothetical or part of pilot initiatives. On the other hand, manufacturers are relying on users to come up with cutting-edge solutions that serve their needs through the use of APIs (application programming interfaces). The rise of APIs is reflected in the number of applications available for wearable devices, including facial recognition, translation, photo manipulation, health monitoring, and social networks. However, developing new applications for each device requires capital and human resources that most libraries can't afford, and in the absence of a culture of innovation and experimentation among librarians and libraries, any custom integration of wearables in libraries will have to come from vendors or other third-party developers. If libraries would consider working together and pooling resources, there might an opportunity for creating an R&D consortium that would be able to focus on building and profiting from this technology.

Privacy and protection of users' data remain the greatest challenges of wearable devices. When Yale libraries introduced Google Glass to its community, privacy concerns related to HIPAA (Health Insurance Portability and Accountability Act) were raised regarding the possibility of the device being used by the medical community in which records or sensitive patient information might be shared (Bruno, 2015). Google Glass and other wearables are not HIPAA-compliant out of the box. Third-party developers are required to build custom operating systems to add an extra layer of security to these devices before they can be used in health care, or in any other situation where privacy and security are of concern. Smart glasses and wearable devices with embedded cameras are also subject to many controversies when used in public spaces (Sultan, 2015). Even before its release, Google Glass was seen as a threat to users' privacy, which fueled strong reactions, including a letter sent by a group of US Congressmen to Google's CEO in June 2013 inquiring about security, data collection, and privacy policy, among other concerns (Mind Commerce, 2014).

Of somewhat lesser concern, but still important to consider, the smaller size of wearable devices tends to impact negatively the processing power, battery capacity, and overall user experience. This can be a problematic issue if the devices are to be used in an extreme environment or for longer periods of time. However, some manufacturers are already offering industrial versions of their consumer devices, such as Vuzix's 3000 series for smart glasses, which targets enterprise users by offering longer battery life, comfortable head mounts, and larger displays to meet their professional needs.

## Conclusion

Mobile devices have the capacity to change how we learn and how we access information. We have just begun to see the opportunities for this technology in the education field as manifested by new policies and approaches adopted to embed users' personal mobile devices into school and workplace. Wearable technologies are a new frontier for educators; we can expect greater impact as the wearables go through different iterations and become increasingly smaller and less intrusive. However, as we have demonstrated in this chapter, education in general, and the library field in particular, don't always offer an attractive return on investment for developers and manufacturers in the same way as health, fitness, or other industries. This lack of financial incentive could impact and delay the benefits this new technology can bring to faculty and students. Therefore, it is important that librarians work at creating opportunities for innovation and experimentation with new devices and technologies to explore their potential in supporting faculty and students as a way of encouraging investment in the technology. Having innovators from the library field with an understanding of the needs and problems that our users are facing will help to ensure the development of adequate solutions more prone to succeed in an educational context, rather than relying on attempts to adapt solutions from other industries to make them fit into our profession. By providing access to these technologies in their early stages, libraries also help to democratize access to information among users of lesser means, increasing their likelihood of being fully engaged with different concepts of connectivity within their learning environment.

Wearables are here, and the future is bright. However, like all technology, they will continue to evolve in ways that are not clear today. At the very least, wearables offer an opportunity for librarians to think about new and improved ways to provide access to their diverse content. Librarians must embrace these technological developments and use every opportunity to experiment and share results. While the future holds great promise, librarians also need to be mindful and proactive in thinking through the still very serious issues surrounding security and privacy. In a world where wearable devices exist and are empowered by sensors and detectors, librarians should also expect to exchange other types of information and data with their users, such as location information, personal information, and physical space and environmental information. These additional streams of information will require new approaches to handle security, safety hazards, copyright, and privacy concerns. Wearables in the academic library will bring neither dystopia nor utopia but rather many exciting new opportunities.

# References

Atlantic, V. (2015). *Sony SmartWear*. Retrieved from http://www.virgin-atlantic.com/us/en/footer/media-centre/press-releases/sony-smartwear/_jcr_content.html.

Bruno, T. (2015). *Wearable technology: Smart watches to Google Glass for libraries*. London: Rowman & Littlefield.

Epson. (n.d.). *Smart Glasses*. Retrieved from https://epson.com/moverio-augmented-reality-smart-glasses?pg=3#sn.

Fleming, G. (2015). *DatadDigest: Announcing our annual benchmark on the state of US consumers and technology in 2015*. Retrieved from http://blogs.forrester.com/gina_fleming/15-09-28-data_digest_announcing_our_annual_benchmark_on_the_state_of_us_consumers_and_technology_in_2015.

Frederick, K. (2015). Dressing for the future: wearable technology. *School Library Monthly, 31*(4), 25–26.

Gartner. (2015). *Gartner's 2015 hype cycle for emerging technologies identifies the computing innovations that organizations should monitor*. Retrieved from http://www.gartner.com/newsroom/id/3114217.

Google. (2015). *We're graduating from Google[x] labs*. Retrieved from https://plus.google.com/+GoogleGlass/posts/9uiwXY42tvc .

Google Glass. (2013). *Explorer story: Andrew Vanden Heuvel [through Google Glass]*. Retrieved from Youtube.com: https://youtu.be/yRrdeFh5-io.

Google Glass. (2016). Retrieved from https://en.wikipedia.org/wiki/Google_Glass.

Irvine, S. o.-U. (2014,). *Another UC Irvine first: Integrating Google Glass into the curriculum*. Retrieved from http://www.som.uci.edu/features/feature-google-glass05142014.asp.

Kim, E. (2015). *Amazon may be working on special smart glasses that let you watch movies*. Retrieved from http://www.businessinsider.com/amazon-smart-glasses-patent-2015-10.

Lee, M. (2013). *Baidu, China's Google, is developing product similar to Google Glass*. (M. Driskill, Editor) Retrieved from http://www.huffingtonpost.com/2013/04/03/baidu-google-glass_n_3004525.html.

LiKamwa, R., Wang, Z., Carroll, A., Xiaozhu Lin, F., & Zhong, L. (2014). *Draining our Glass: an energy and heat characterization of Google Glass*. Retrieved from http://www.ruf.rice.edu/~mobile/publications/likamwa2014glass.pdf.

Mind Commerce. (2014). *Wearable technology in industry verticals 2014–2019*. Mind Commerce Publishing.

Moline, N. (2015, 12, 29). Google Glass questions. (A. Ajmi, Interviewer).

Nichol, P. B. (2015). *Wearables 2.0: The future Google Glass to UberNurse*. Retrieved from https://www.linkedin.com/pulse/wearables-20-future-googleglass-ubernurse-peter-b-nichol.

OET Exhibits List. (2016). Retrieved from: https://apps.fcc.gov/oetcf/eas/reports/ViewExhibitReport.cfm?mode=Exhibits&RequestTimeout=500&calledFromFrame=N&ap-

plication_id=eDyH1HI%2FRcK9NnzZ4ggP6w%3D%3D&fcc_id=A4R-GG1.
Rainie, L. (2016). Libraries & perpetual learning. *Computers in Libraries Conference*. Washington, DC, USA.
Riskin, R. (2016, 1, 13). Google Glass article from Roxann 2016. (A. Ajmi, Interviewer).
SmartEyeglass. (n.d.). Retrieved from https://developer.sony.com/devices/mobile-accessories/smarteyeglass/.
Sony. (2015). *Single-Lens Display Module demo: The concept model SmartEyeglass Attach! [video]*. Retrieved from https://developer.sony.com/2015/02/12/single-lens-display-module-demo-smarteyeglass-attach/.
Sultan, N. (2015). Reflective thoughts on the potential and challenges of wearable technology for healthcare provision and medical education. *International Journal of Information Management, 35*(5), 521–526.
Vuzix. (n.d.). *3000 Series Smart Glasses*. Retrieved from https://www.vuzix.com/Products/Series-3000-Smart-Glasses.
Waller, J. (2016, 1, 16). Google Glass question. (A. Ajmi, Interviewer).
Wikipedia. (2016). *Pebble (watch)*. Retrieved from https://en.wikipedia.org/wiki/Pebble_%28watch%29.
Winkle, A. (2015, 12, 28). Google Glass question. (A. Ajmi, Interviewer).

# About the Authors

## About the Editors

**Robin Canuel** is an associate librarian at McGill University and currently serves as Head of the Humanities and Social Sciences Library at McGill. He has coauthored several articles on a variety of topics, including tailoring information literacy instruction to specific constituencies, the design and use of mobile websites for academic libraries, using tablets for teaching and research, and leveraging mobile apps in academic libraries. He earned a BA from McGill University in 2000, and his MLIS in 2002, also from McGill.

**Chad Crichton** served as Coordinator of Reference, Research and Instruction at the University of Toronto Scarborough for eight years. He is currently the campus Liaison Librarian for both the Department of English and the Department of Arts, Culture, and Media. He has presented on the topic of mobile technology at library conferences in both North America and Europe and has also published a number of scholarly articles on the topic. Chad earned an Honours BA from Queen's University in 1998, an MA in English literature from Wilfrid Laurier University in 1999, and his MLIS from McGill University in 2002.

## About the Authors

**Ayyoub Ajmi** (@ayyoovod) is a Digital Communications and Learning Initiatives Librarian at the University of Missouri–Kansas City School of Law. He is building and managing an integrated digital communications platform, which provides access to the law school library and its digital resources, supports law faculty's effective use of technology to enhance student learning, and facilitates information and communication among various constituencies of the law school.

**Edward Bilodeau** is the management librarian at the McGill University Library. He has extensive experience in both academia and the private sector, primarily in the areas of strategy, information architecture, communications, and project and team management. His current interests include discovery systems, social media, and implementing scalable and sustainable library practices.

# ABOUT THE AUTHORS

**Nathan E. Carlson** is an assistant professor and librarian at Metropolitan State University in St. Paul, Minnesota. He teaches credit-bearing information literacy classes and serves on the e-resources and web teams for the library. His work centers on improving the user experience of the Web, databases, and discovery.

**Jennifer DeJonghe** is a librarian and professor at Metropolitan State University in St. Paul, Minnesota. She teaches credit-bearing courses on information literacy and social media and is a member of the library web and UX team. She also works to integrate emerging technologies into library services and online instruction.

**Barry Falls** is the Roaming Services Coordinator for J. Murrey Atkins Library at UNC Charlotte, where he has worked for five years. He graduated from UNC Charlotte with a bachelor of arts degree in sociology with a minor in journalism. Currently, Barry is working toward a master of library and information studies degree from UNC Greensboro, where his studies are focused on technology in academic libraries.

**Tiana Faultry-Okonkwo** is the Visitors and Tours Coordinator at Texas A&M University Libraries in College Station, Texas. She is also the coordinator of the libraries' instruction classes for high school groups in Bryan—College Station and surrounding communities.

**Stacy Gilbert** is a Business Librarian at Texas A&M University. She received a master of science in library science degree from the University of North Carolina at Chapel Hill and a master of science in media and communications from the London School of Economics and Political Science.

**Stephanie Graves** is an associate professor and Director of Learning and Outreach at Texas A&M University Libraries. Her research explores the intersection of information literacy, user experience, reference, and emerging technologies. She is active in ALA and Reference and Users Services Association (RUSA).

**Jim Hahn** is Orientation Services & Environments Librarian and Associate Professor at the University Library at the University of Illinois. His duties include developing and evaluating prototype technologies that focus on enabling undergraduates to discover library resources and services that support learning and research and to integrate them into their work.

**Yusuke Ishimura** has an MLIS degree from Dalhousie University and PhD from McGill University in Canada. He was previously a faculty member in the Information Science program at Edith Cowan University, Australia, before moving to his current position as an Assistant Professor at Tokyo City University, Japan. His current research focuses on the integration of information literacy and emerging technologies.

## About the Authors

**Keven Jeffery** is the Digital Technologies Librarian at San Diego State University. He received his MLIS from the University of Western Ontario in 2001 and has been applying new technologies to the provision of library services ever since. He has presented nationally on technology projects and has published in journals such as *Library Hi Tech* and *The Journal of Web Librarianship*.

**Wayne Johnston** has been a librarian at the University of Guelph Library since 2004. His work has focused on supporting research, especially through the use of emerging technologies. His fieldwork has included projects in Ghana, Nepal, Bolivia, Croatia, and Canada's Arctic.

**Maureen (Molly) Knapp** is a Training Development Specialist for the US National Network of Libraries of Medicine. She was previously employed at Tulane University's Matas Library of the Health Sciences. A librarian with fifteen years of experience in academic medical libraries, she is currently living the dream in Houston, Texas.

**Avery Le** holds a BA from the University of Southern California and an MLIS from Florida State University. She also holds a JD from the University of Florida Levin College of Law, where she currently serves as the Technology and Digital Services Librarian and adjunct professor of legal research.

**Mê-Linh Lê**, MA, MLIS, AHIP, is the Health Sciences Centre Librarian at the University of Manitoba. She has long been interested in emerging technologies and how librarians can use them in their professional practice. Her work in this area has been published in Canadian, American, and international journals.

**Yoo Young Lee** is the Digital User Experience Librarian and Liaison Librarian to the School of Health and Rehabilitation Sciences at IUPUI. Her interests include the relationship between library instruction and UX, user-centered design, emerging technologies, and analytical tools. She earned her MLIS degree from McGill University.

**Sarah LeMire** is the First Year Experience and Outreach Librarian at Texas A&M University. She is interested in information literacy instruction, assessment, scalability of instruction and outreach, and outreach to special populations, especially student veterans.

**Teresa Maceira** is currently the Acting Co-Head of Reference, Outreach, and Instruction at the University of Massachusetts Boston. She also serves as the liaison librarian to the social sciences. Teresa received a master of science in library information Sciences from Simmons College in 2011.

**Beth Martin** is the Head of Assessment for J. Murrey Atkins Library at UNC Charlotte. She is also working on a PhD in educational research and policy analysis at North Carolina State University. Prior to her work in libraries, Beth worked in information technology with a focus on network infrastructure design and implementation for businesses such as Lucent Technologies and eMusic.com.

**Dr. Martin Masek** is a Senior Lecturer and leader of the Transformational Games Research Group at Edith Cowan University. His research interests include the application of computer game technology to motivate learning and exercise and for simulation and monitoring.

**Willie Miller** is the Liaison Librarian to the School of Informatics & Computing and the Department of Journalism & Public Relations at IUPUI. His research interests include library instruction, instructional technology, learning environments, and student engagement. He earned his MLS from Indiana University.

**Abby Moore** is the interim Head of Research and Instruction Services as well as an Education Librarian at UNC Charlotte. Prior to her work at UNC Charlotte, Abby was the Education Librarian at University of South Dakota. Her background is in K–12 education, specifically high school English education and school libraries.

**Jordan Nielsen** is the Business Data Librarian at San Diego State University. He received an MS in information sciences from the University of Tennessee in 2012. He regularly conducts presentations about technology and its impact on teaching and learning in libraries, and he has published in the *Journal of Library and Information Services in Distance Learning* and the *Qualitative and Quantitative Methods in Libraries Journal*.

**Caitlin A. Pike** is the Liaison Librarian to the School of Nursing and the Medical Humanities and Health Studies program at IUPUI. Her research interests include open access, mobile technology, and developing relationships with faculty to better integrate library services. She received her MLS from North Carolina Central University.

**Hailie Posey** is the Digital Publishing Services Coordinator at Providence College's Phillips Memorial Library. She earned her BA in Spanish and anthropology from Dickinson College and her MLIS, along with an MA in history, from the University of Rhode Island. Hailie's research interests include open access and scholarly communication, open educational resources, mobile technologies, digital humanities, and digital literacy.

# About the Authors

**Michael Robak** received his MSLIS from the University of Illinois after practicing law for fifteen years. Michael was then Director of Legal Research for Charles River Associates International. In 2008, Michael moved to academic law librarianship at the University of Illinois College of Law. He joined the University of Missouri–Kansas City School of Law as Associate Law Library Director and Chief Technology Officer in 2011.

**Regina Lee Roberts** is a Collection Development Librarian in the Humanities and Social Sciences at Stanford University Libraries. Roberts designs learner-centered library workshops infused with course-specific content and research design strategies. Her subject area responsibilities include Anthropology & Archaeology, Communication & Journalism, Feminist Studies, Lusophone Africa, and Sociology.

**Alec Sonsteby**, MS, MBA, is a librarian and associate professor at Metropolitan State University, St. Paul, Minnesota. There he is a member of the library's web team, provides reference services, and teaches credit-bearing information literacy courses. He received his library degree from the University of Illinois at Urbana-Champaign.

**Mattie Taormina**, director of the Sutro Library in San Francisco, was the head of public services for Stanford Special Collections from 2006 to 2016. She is the coeditor of *Using Primary Sources: Hands-On Instructional Exercises* and the 2016 President of the Society of California Archivists.

**Junior Tidal** is the Multimedia and Web Services Librarian, Associate Professor for the Ursula C. Schwerin Library at New York City College of Technology, CUNY. The author of *Usability and the Mobile Web: A LITA Guide*, he has also published in the *Journal of Web Librarianship*, *Computers in Libraries*, and *code4Lib Journal*.

**Danitta Wong** is currently the Acting Co-Head of Reference, Outreach, and Instruction at the University of Massachusetts Boston. She also serves as the liaison librarian for the College of Science and Mathematics. Danitta received a master of science in library information science from Simmons College in 2010.